The War of the

of the

and

Fishes

The War of the JESUS of the and DARWIN Fishes

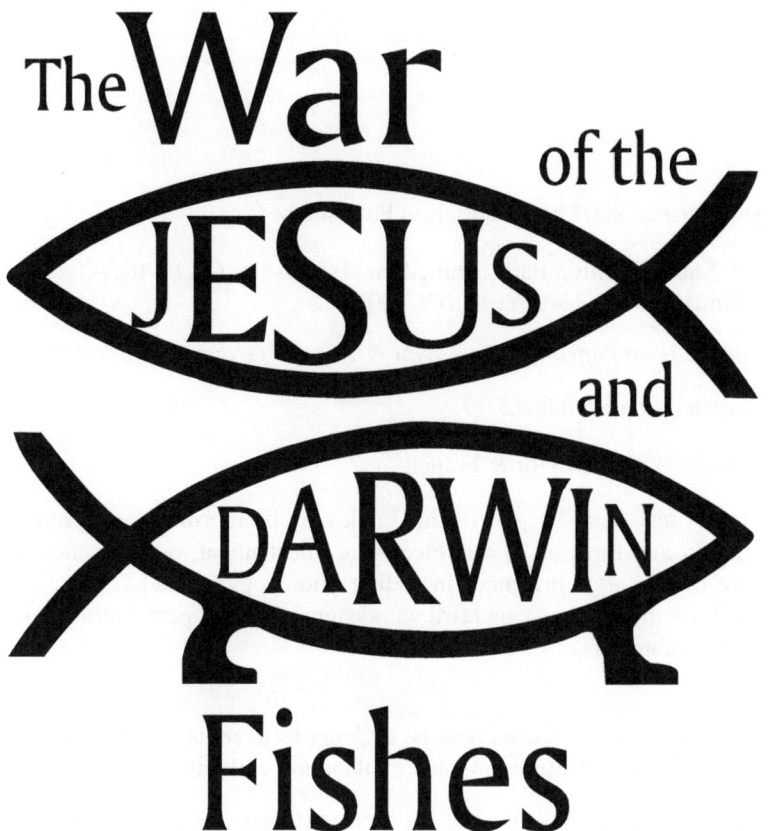

Religion and Science in the Postmodern World

John C. Caiazza

Routledge
Taylor & Francis Group

LONDON AND NEW YORK

First published 2007 by Transaction Publishers

2 Park Square, Milton Park, Abingdon, Oxfordshire OX14 4RN
711 Third Avenue, New York, NY 10017

Routledge is an imprint of the Taylor & Francis Group, an informa business

First issued in paperback 2017

Library of Congress Catalog Number: 2007017001

Library of Congress Cataloging-in-Publication Data

Caiazza, John.
 The war of the Jesus and Darwin fishes : religion and science in the postmodern world / John C. Caiazza.
    p. cm.
 Includes bibliographical references and index.
 ISBN 978-0-7658-0380-1 (alk. paper)
    I. Title.

BL240.3.C334 2007
201'.65—dc22   2007017001

ISBN 13: 978-0-7658-0380-1 (hbk)
ISBN 13: 978-1-138-51712-7 (pbk)

Dedication

To the memory of my parents, Nicholas A. and Marion C. Caiazza,
"Nick and Marion"

Contents

Acknowledgements

Books, while seemingly the fruit of solitary effort when sitting on the shelf or in a reader's hand, are in fact the product of many hands, and often what seems to be the insight and perspicacity of the author are in fact the borrowed finery taken from the minds of many others. A look at the bibliography will indicate where most if not all of my ideas and insights came from.

Conversations with many people helped to distill the ideas in this book, including conversations on the topic of religion and culture with two brothers-in-law: Mike Hickey, M.Div., and Steve Bell. In the summer of 2005 I happily attended a seminar supported by the National Endowment for the Humanities held at Boston College on the subject "Religious Diversity and the Common Good." The seminar discussions, which led to productive insights, took place with an exceptional group of scholars of religion from varied backgrounds and were excellently led by Professor Alan Wolfe.

I wish to acknowledge the continued interest and advice given by the editors of several journals including Philip Hefner of *Zygon*, George Carey of *The Political Science Reviewer,* and especially George Panichas long time editor of *Modern Age,* the journal where conservatives come to think. Also, I want to acknowledge the patience and help of Irving Louis Horowitz, the Editorial Director of Transaction Publishers, who perused and evaluated two previous versions of the present work and approved this version.

I wish to thank my colleagues at Rivier College for their encouragement and support, especially Professor Herman Tavani of the Philosophy Department and Dean of the College Albert DeCiccio. Chapter 6 originally appeared in the Rivier College on-line journal edited by Professor Vladimir Riabov, and my thanks to him.

My son, Paul, gave technical assistance with the PC and laptop computers, providing an instance of how the younger generation grasps the new electronic media much more rapidly than the older. Credit given by authors to spouses and lovers is usually of the general kind—"to Titania for everything" or "to Guilliame who knows why"—but in this case I can be more specific. My wife, Jo, encouraged me to buy the PC and laptop on which the manuscript was typed, set up a space for me to do my writing, and provided the necessary support that let me pursue a supplementary career of writing and studying when I could have been out earning money.

Introduction: The War of the Jesus and Darwin Fishes

A variety of stickers and plaques stuck onto the rear ends of automobiles is a constant feature of driving in America. Everyone has an opinion, it seems, and the driver just ahead of you seems compelled to advertise it on the back end of his or her personal transportation. Besides the political bumper stickers and the brand marks for auto dealerships, you will sometimes see the outline image of a fish. If you are attentive to the fishes as they occasionally appear on the car ahead, you will discover that they come in two distinct varieties, one which is a bare outline of a fish or the outline with a cross or the name "JESUS" inscribed within, the other variety has little legs attached to the underside with the name "DARWIN" or the term "EVOLUTION" inscribed within. It is a noteworthy phenomenon that has attracted the attention of serious commentators and social scientists and is an indicator of a new phase in an old cultural conflict. It is a conflict which extends back at least to Victorian times 150 years ago, but to the historically minded it extends back to the time of the Enlightenment. Perhaps even beyond that, for the Jesus and Darwin fishes are a symbol that the "war" between religion and science is still alive in Western culture.[1]

The war between the Jesus and Darwin fishes started when contemporary Christians began advertising their belief by means of the simple fish outline. The fish in ancient times was used as a symbol of Christian belief especially during persecution as a means of Christians recognizing one another without giving themselves away. (The word "fish" in Greek is an acronym for "Jesus Christ, Son of God and Savior.") On the rear of an automobile the fish outline serves the same function, but the renewed popularity of the Jesus fish in itself did not constitute a conflict. The conflict began when advocates of a scientific point of view, or people who were simply put off by public affirmations of religious affiliation, responded to the Jesus fish with the Darwinian version as a public statement of their own persuasion. When first observed, the phenomenon of dueling fish symbols seemed to have a humorous aspect and no more significance than the bumper stickers that say "my child was student of the month" countered by "my kid beat up your student of the month." Sociologists might perceive deep meanings consistent with class or

1

income differentials, but the Jesus and Darwin fishes combat could be perceived as nothing more than an expression of healthy, rough American humor. However, the Darwin fishes are not intended as gentle mockery but as direct opposition to the Jesus fishes and the duel has a rather nasty aspect, as if the Darwinian fishes are in mortal combat against perceived intrusions of bigotry and theocracy signified by the appearance of the Jesus fishes. In short, the war between the two fish symbols has an underlying aspect that is quite serious. But if the Darwin fishes were in response to the Jesus fishes, then what was the cause of the appearance of the Jesus fishes in the first place?

The cause of a social phenomenon such as the Jesus fishes is not always readily discernible, much like fashions come and go and are often not related to anything but themselves (long skirts, baggy shorts, etc.). The Jesus fishes do have a discernible cause, however, namely the sense that many traditional minded Christians and believers of virtually all other religions in the United States have that the ambient culture has become so materialistic and degraded as to bear comparison with no other time in American history than the debauched times of the end of the Roman Empire. (One of the important causes of Muslim anger with Western culture, that provokes disdain and fear, is the blatant sexualization of our popular culture and our lax sexual morality.) Traditional minded religious believers are further aware that the culture has become explicitly anti-Christian and antireligious. At a time when public policy in the areas of abortion and marriage have contradicted traditional norms, many Christians have begun to feel that they are aliens in their own culture. Like the Christians of ancient Rome, they feel they are subject to a persecution that, if it does not make them criminals on a cultural level, attempts to suppress the full expression of their faith by means of ridicule and contempt. Thus, the appearance of the Jesus fishes on the rear ends of automobiles.

As already indicated, the appearance of the Darwin fishes is also a fear based phenomenon, namely, the fear that a potent combination of right-wing political organization and evangelical religious sentiment now constitutes a danger to the separation of church and state. More exactly, the fear is that antiprogressive forces from the southern and middle west of the nation (the "bible belt" and flyover country which are the parts of the nation largely unsympathetic to Liberalism) will reverse the recent advances in the rights of women, gays, and ethnic minorities since the new "religious right" has (as it always has had) the will and now the political power to "turn back the clock." This new religious movement cannot be ignored, and its power now makes it obligatory to face down its proponents. As it happens, the issue of evolution is a major flash point; the advocates of a new theocratic order are using new versions of the antiquated argument from design for the existence of God to force school boards to vote that "intelligent design" be taught in public schools alongside the theory of evolution. Thus, the need to oppose any public display of a Christian symbol, even a seemingly inoffensive one such as the wordless Jesus fish outline on the rear of automobiles.

Since the original debates over evolution were carried out between Anglican divines and Victorian naturalists in learned articles and large tomes of intricate argumentations, the war between the Jesus and the Darwin fishes seems to suggest the ability of American culture to vulgarize anything it engages. Yet the bumper sticker war has a social subtext of some intensity and signifies a serious contemporary divide in American culture on two levels. On the political level, the argument reflects in broad terms the present conflict in American politics between "red states" and "blue states"— rival political alignments. And on an intellectual level, the war reflects the ideological tension at the basis of modern Western culture since the Enlightenment—the conflict between religion and science. It is on this point that the most fundamental aspect of the war between the Jesus and the Darwin fishes takes place and serves as an introduction into an old controversy which now has taken on a new form.

The evolution controversy is the best known and currently most compelling aspect of the conflict between religion and science, but, as indicated, the evolution controversy itself has an older provenance extending back to the time of the Enlightenment (or according to another account we will meet later, back to the third century A.D.). The Galileo case may be used, however, as the point at which the conflict arises because it was at that time that the Medieval worldview (expressed in art and popular culture as much as in the grand philosophical and theological systems of the time) was rent asunder by the intrusion of new empirical discoveries and a new scientific method of discerning ultimate truths about the universe.[2] From that point forward in Western history it has seemed as if the serious, educated person had to choose between a rational and testable but emotionally and spiritually vacant understanding of the universe, or a humanistic and spiritual understanding which was rationally indefensible. The religion/science conflict has gone many rounds since then, from the French Enlightenment in the eighteenth century, to the positivism of Comte in the nineteenth, to the positivism of the Vienna Circle in the twentieth, to the reductive accounts of religious beliefs and the assertive scientific materialism of contemporary commentators such as Dennett and Dawkins.

The bumper sticker war of the Jesus and Darwin fishes serves to bring to mind that the conflict between science and religion now takes place in a new cultural environment. For better or for worse, I will call this new cultural environment "postmodern" rather than "global," "international," or "cosmopolitan." While these other terms have the advantage of referring to the reality that the human race is now in the process of erecting a new culture which extends worldwide, they also evade the more ideological and intellectual consequences of the culture now in the process of being formed. This understanding that a new culture is being formed brings a new perspective and complexity not seen in prior iterations of the debate between religion and science, for postmodern culture has downgraded the importance of both religion and science. Precisely because both science and religion were necessary parts of western culture they are both attacked, ignored, and often dismissed, but in this book I will argue that both are necessary if the evidently unsatisfactory and destructive aspects of the postmodern world are to be overcome.

Notes

1.   Websites where the Jesus and Darwin fish stickers are advertised are, respectively, www.planeticthus.com which sells Christian auto accessories and www. prankplace.com which sells "pranks, gags … and practical jokes."
2.   Ian G. Barbour, *Religion and Science* (San Francisco, CA: HarperCollins, 1997), 3–17.

Part I

Religion and Science in the Postmodern World

1

The Postmodern World

Postmodernism is a True Worldview

To speak of a "postmodern world" is to relate how an explicitly drawn philosophy reflects the way the world currently is, and the world today is now more than ever one world as distinctions between nations and among peoples, histories, and traditions are breached with constancy and regularity. The usual forces are acknowledged here; the transport of goods, commerce, travel, communication, finance, education, and even labor markets have become internationalized. In this international environment, the "noosphere" of which Teilhard wrote in mystical and poetic terms, is realized by means of electronic communication via digital technology and satellite. So that, for example, a ferry disaster in Bangladesh takes place in real time on our full color HDTV sets as we watch the bodies floating up on shore and hear the weeping of the relatives of the victims, or the decision of a Japanese investor is made actual so that hundreds of thousands of dollars are transferred virtually instantaneously from a bank in Tokyo to the New York Stock Exchange by means of several keystrokes on a laptop. The ubiquity and efficiency of current electronic communication allows us to directly perceive all the many cultures throughout the world, Eastern and Western (although it is getting harder to tell them apart), modern and primitive (although the distinction is usually left unstated), contiguous with one another as they appear on television, computer, and ipod screens. Thus, there is the theme of multiculturalism in the postmodern word in which each culture commands respect equally without distinction, the constant succession of cultural images acting as a spur to tolerance.

In this international environment, the personal belief systems of the participants become irrelevant, if not to the individuals involved certainly to the processes of international intercourse. It does not matter whether an individual believes that Jesus rose from the dead, that Mohammed was the last of the prophets, or that there is no deity at all; whether she believes that illness is caused by wicked spirits, an unhealthy frame of mind, or by viruses. The fact is that the virtues involved in living in an internationalized environment are not those which promote personal or interior virtues but ones which exclusively refer to how we as persons relate to each other in this new environment. Here the virtues are those of tolerance and mutual recognition of the commonality of mankind, which in turn promote the acceptance of all the representatives of the human race and the suppression of any sense of differentness that may appear when we meet others. As in the days of the Roman Empire serious and reflective people understand that there is a new world thing, not a new world order exactly, but a commonality that can be observed and described in moral terms. Martha Nussbaum and Anthony Appiah discourse in the same manner, therefore, as did the Cynics of ancient Greece and the Stoics of ancient Rome, about persons being citizens not of their nations but of the world; nothing human is alien to them, they might as well say as did the Roman philosopher.[1] The new Stoics quote the writings of the Roman emperor Marcus Aurelius and promote a philosophy of cosmopolitanism that is a vision of the universal city of man, or, in the gender neutral language of cosmopolitanism, the human city.

In this environment, assertions that a particular tradition or "narrative" holds a special place (i.e., is *true* in the sense that whatever competing accounts may state that contradict it are false) are not merely discouraged but put out of consideration as a matter of principle—a point made by the postmodern theorist Derrida.[2] The argument does not take place on the question of what evidence, for example, there is that Jesus rose from the dead as that implies an assertion that the Christian religion is true and that Islam and Hinduism are false. Such assertions are simply not to be considered by moral people in the postmodern environment. Likewise, open debate about the superiority of scientific method over faith, superstition, intuition, etc. are discouraged; one result of which is the equalization of medical treatment so that the incantations of shamans, the homeopathic use of herbs, or acupuncture therapy are serious options for medical treatment. The scientific basis of Western medical knowledge has been transmogrified from a guarantor of its effectiveness to the belief that scientific medicine is itself a form of mockery of non-Western cultures and non-European peoples, as argued by the postmodern historian/philosopher Foucault.[3]

The postmodern worldview does not tend to assert the equality of all races, genders, cultures, and histories, but to favor those that have been the victim of oppression including people of color, women, homosexuals, and non-Western cultures such as Eastern and American Indian. It also tends to elevate the histories of anything other than that of male political leadership, i.e., the histories of women, workers, and peoples formerly subject to Western colonial rule. The postmodern ideals of inclusion and tolerance have a direct political impact which is manifestly left-wing or progressive and is involved with a deep and unrelenting sense of resentment and moral self-justification. There is a pervasive sense that identifying with the oppressed yields a moral superiority to the postmodern critic and lends immediate credence to the criticisms and accusations made against those individuals or social entities which are white, male, heterosexual, Western, and European-American. Hence, the stance taken in the postmodern worldview tends to be judgmental and critical to the point that the objects of critical review are not given initial credence; this means, in turn, that the usual result of postmodern criticism does not have the flavor or form of debate but of prosecutorial accusation. In this way, there is an unfortunate and ironic tendency for postmodern discourse to praise the virtue of tolerance while manifesting an attitude of intolerance. The sense seems to be that white men, Europeans, and other traditional authorities have had it their own way for so long in the power relations between themselves and all the categories of the historically oppressed that their point of view no longer needs or deserves to be heard. And included among oppressive points of view disdained by the postmodern sensibility, are the accounts of both the religious and scientific heritages of Western civilization.

The Philosophy of Postmodernism

Postmodernism as an attitude is a reflection of the development of the reality of a worldwide culture, but the distinctive aspect of the human species is that it can reflect on its circumstances and understand them. Thus, a serious and explicit intellectual underpinning is discernible in the postmodern worldview—an actual philosophy with its own epistemology, metaphysics, ethical theory, and aesthetic manner. The general postmodern attitude is well expressed by philosopher Richard Rorty's use of the term "ironism," which on one level denies that truth is obtainable by any known means and on another level is purely relative to social circumstances; so, declarations of truth reflect only the degree and type of power of those institutions and persons who enunciate a vision of truth. At the same time, postmodernism avows a set of universal values which aim to defend the rights of individuals, include the perspectives of all the cultures across the world, and be respectful of all classes, races, genders, and ethnicities especially those which have suffered oppression from Western colonialism and patriarchy. The irony resides in the fact that, despite the contradiction between epistemic and moral relativism on one hand and the insistence of a scheme of universal values on the other, postmodernism insists on the priority of its own set of values. Rorty puts the seemingly self-contradictory situation of current Liberalism in a philosophically sophisticated way, citing among others Habermas and Dewey.[4] There is an unrecognized irony in the postmodern attitude, however, since while it encourages different and dissident points of view, by also rejecting all "master narratives" or "meta-narratives," it excludes the truth claims of modern science and the Jewish and Christian religions.

Epistemology.   The epistemology of postmodernism is explicitly relativistic and begins almost in a formal sense in Rorty's major work, *Philosophy and The Mirror of Nature*, a long and influential disquisition on the most important philosopher of the modern age, Rene Descartes. Rorty interprets Descartes, as the paradigmatic representative of the scientific view, writting that there is a permanent, physical reality which the human intellect can accurately represent by means of experiment and mathematical laws. Rorty proceeds at length to attack the notion that scientific research elicits objective truth and is the paradigm of rationality. By contrast he states, "we see knowledge as a matter of conversation and of social practice rather than as an attempt to mirror nature," and the "crucial premise of this argument is that we understand knowledge when we understand the social justification of belief, and thus have no need to view it as accuracy of representation."[5] The effect of postmodernism's relativistic epistemology lies in its application to the social sphere where it maintains that each person or social tradition has its own truth, thus denying the possibility of the existence of any universal truths. In this way, the validity of canonical histories and the implied superiority of the characters and causes described in them are thereby denied as a whole. Washington and Jefferson, in this view, are not heroes of the American story of freedom but rather slaveholders who acted to preserve the power of rich, white males.

Metaphysics.   Oddly perhaps, the political and cultural implications of postmodern philosophy become apparent in an attempt to discern its metaphysics, i.e., its theory of reality. In contemporary critical references to "essentialism" and positive references to "anti-essentialism" lies the evidence that postmodern metaphysics denies that there is a knowable external reality beyond perception. "The ironist, by contrast, is a nominalist and a historicist. She thinks nothing has an intrinsic nature, a real essence."[6] In effect, everything is noumenal, and reality is a "construct." The validity of the notion of reality is denied, understood as an external presence which ultimately structures the shape and behavior of individual human beings and of society. Postmodernism promotes the idea of social constructivism, that so-called "realities" are merely the agreed upon versions enforced by particular societies or the dominant elements within it. In cultural and political debates, the metaphysics of anti-essentialism can be useful, particularly in conflicts regarding the role of women where traditionalist arguments that women's biology and emotional structure is, in effect, their destiny can be countered by arguing that women's roles are social constructs maintained by the patriarchal power structure and, thus, are not the result of nature or an aspect of God's plan. This postmodernist manner of approaching cultural disagreements discourages case by case analysis of whether, for instance, single mothers are as successful in raising their children as married ones. It relies instead on expressions of moral indignation which implies that raising such a question is an insult to single mothers and a denigration of women's equal status in society.

Ethics. Postmodern ethics is concerned with the relations between people and among cultures and its primary value in this regard is tolerance, that is, the full acceptance of individuals as members of groups to which no negative stereotype or prejudicial characteristic is attached. Great emphasis is put upon eliminating negative prejudices toward racial and religious minorities, women, and people who engage in nontraditional sexual lifestyles. Great sympathy is manifested toward victims of oppression, although less as individuals than as members of a class or group. Postmodern politics tends to be left wing or progressive and often refers to "the whole world" or "the international community" as the ultimate basis of ethical concern as opposed to one's nation, religion, or family.

On a personal level, postmodern ethics in seeking the good of the individual is concerned mainly with health and good living; it discourages such habits as cigarette smoking and whisky drinking, which were formerly seen as permissible activities, while promoting healthy diet, regular medical examinations, exercise, and weight loss. These ethics of healthy living extend to the emotions as well as the body, and great attention is paid to psychological adjustment and balance as people are instructed how to control their appetites, overcome negative attitudes, and survive personal crises with the aid of counseling by professionals trained to provide guidance to individuals and families. The ethics of postmodernism is as relativistic as its epistemology, as no one rule or moral law is given precedence over any other; the legitimacy of social entities including the state, religion, civil authority, parents, and the family unit are, in effect, denied or treated with immediate suspicion. The idea of a moral law implanted as an element of nature developed in the West by the Psalmist, Socrates, Plato, Cicero, St. Paul, the Stoics, Thomas Aquinas, Grotius, Locke, Jefferson, and Lincoln is discounted and given no credence. The only element selected out by postmodern ethics from the patrimony of the West is its regard for individual rights. Postmodern ethics is, in a strict sense, antinomian.

Aesthetic.   Finally, the total attitude or aesthetic of postmodernism elevates personal and dissident visions above those which are canonical or claim to represent reality. This dissident attitude is reflected in the current popularity of the literature of personal memoirs, alternate histories, and the dreamlike narratives of authors from South America such as Borges and Marquez. Movie, television, and computer games utilize science fiction, which has emerged in the last forty years from a closeted literary genre to wide, general influence within popular culture. Science fiction and fantasy feature alternate universes galore, among these stand out the dark fantasies of the late Philip K. Dick in *Blade Runner* and *The Matrix.* In these movies, the commonly accepted reality is pictured as manufactured by malign powers which the heroes resist not only by force of arms but by persistent and energetic creation of their own counter-reality. In these fantasies, another feature of the postmodern aesthetic finds expression, namely, the permanent sense of aggrievement; the sense that resistance to order is a moral imperative. The postmodern sensibility disdains essences and thereby luxuriates in a dissociative attitude in which connections and mutual responses are refused recognition. For in denying essences, it denies that there are inherent connections between events, actions, and things and not merely accepts but embraces the consequence of universal incoherence. Such an attitude necessarily disdains the ultimate visions of religion and science, as the purpose of both in a philosophic sense is to render the universe comprehensible in its details according to basic laws understandable by all human beings.

Rearticulation of the Roles of Religion and Science

The postmodern cultural environment is, by definition and in fact, different from the modern (or pre-postmodern) environment which predominated until about 1960. The most important difference for matters of faith and reason is that nowadays traditional religion and modern science are not given the same degree of credibility that they previously possessed in Western culture. While the idea of a conflict between religion and science is a nineteenth-century idea, it persists into the twenty-first and necessarily takes on a different form effected by the change in the culture. Namely, the shift from a strong sense of confidence in Western civilization typical of the Victorian age to one of demoralization produced by a full century of horribly destructive wars—wars made more deadly and destructive by means of advanced technology and motivated by totalitarian ideology. The religious vision of the peaceable kingdom was not able to measurably inhibit either World War I or World War II, and, in fact, the religious identity of national populations was used by governments to sustain morale during wartime as the Kaiser appealed to the Lutheran church and Stalin to the Russian Orthodox church. Science was turned from the ideal of reason in Western civilization to scientific technology which provided such awful weapons as machine guns, poison gas, and enhanced chemical explosives. Having provided its emotional and physical structure, once modern Western civilization collapsed into demonically chaotic destruction, religion and science came under suspicion and attack along with the other traditional institutions.

Despite the postmodern world's disparagement of science and religion, it does not dismiss them; on the contrary it finds specific uses for them, albeit in a manner which tends to trivialize them. Third world economic development, for example, is impossible without the use of scientific technologies to increase crop yields or to detect global social trends which require the use of scientific sampling techniques. Science is reduced from knowledge of certain truths about the physical universe to a technique for social progress. Religion catches the attention of the worldwide electronic communications system but mainly as a source of picturesque spectacle: an elephant run in Hindu India, a convocation of red-robed cardinals in Rome, worship dances among tropical tribes. Religion is reduced from an expression of mankind's search for the divine to mere spectacle; the inner and most compelling aspects of religious belief are ignored because they are not innately pictorial.

The trends run in both directions however; from the postmodern world to religion and science, and from science and religion to the postmodern cultural environment. There have been substantial changes within religion and science which have themselves been a cause of the change in Western civilization from a modern to a postmodern outlook. Religion in the West, mainly the Christian religion, has had to face the challenge of diversity and conflict by exposure to Eastern religions and its own history of anti-Semitism, and it has attempted to resolve conflicts among Protestants, Catholics, and Orthodox. This has had a double effect—one positive and another negative. Positively, it has led to a new era of ecumenism and attempts at mutual understanding that were absent before; representatives of different religions not only meet to have discussions at international convocations but more frequently co-operate in local social missions. Negatively, the new diversity has led to a weakening of truth claims of religion generally, and it has become more difficult to promote the exclusive truth of a particular religious tradition without appearing arrogant.

Science as well as religion has come under direct attack in the postmodern world because of its assertion, defined originally by Francis Bacon, that the scientific method is a sure way to obtain truth conflicts with the epistemological relativism which is an essential element of the postmodern worldview. Also, it is scientific technology which provided the means of Western dominance over virtually the entire world starting from the eighteenth century with the rise of the British and Napoleonic empires. Therefore, science is seen, not as a morally neutral agent whose application can harm as well as benefit mankind, but as an agent of colonial and patriarchal domination. As such it has come under ferocious attack by multicultural and feminist critics to the point where advocates of science have responded in force.[7] Interestingly, some theologians and philosophers of religion have taken up epistemological relativism as an intellectual means of degrading science and the notion of objective truth, hoping to establish religious belief as on a logical par with scientific conviction. However, this response also negates the possibility of establishing that religion has a rational basis and makes it appear that religious ethical commands are simply arbitrary or that the notion of God does not designate a reality but a mere choice among possible worldviews. Thus, it has been said that "epistemological relativism leads to moral chaos."

The conflict between religion and science in the postmodern world, thus, has complexities and an edginess not previously seen. It appears that there are some points of mutual vulnerability in that both must deny the relativity of truth if either is to have its validity assured. But just as religion has evolved in a postmodern fashion to recognize diversity, so also must the content of certain scientific theories and the methodology of some scientific fields which have postmodern implications. Indeed, quantum mechanics with its uncertainty principle and subjectivist implication for human knowledge has become a major cultural element of the postmodern sensibility. Although much of popular writing about contemporary science continues to reinforce its traditional link to materialism and naturalism (especially in writings about evolution), writing about contemporary physics and cosmology seem to have the opposite thrust. They are opening up vistas, long thought to be closed by science, of a universe manifesting inherent purpose, a cosmos designed to provide a home for the human race, and even pointing to the reality of God.[8] What the evolutionary biologists seem to be taking away from religion, the physicists seem to be putting back.

One aspect of the postmodern context is the general idea that such things as gender roles or scientific laws are not real in themselves but instead are social constructs which are, in effect, artificial, intellectual, and social structures. The conflict between science and religion can be seen in the same light, i.e., not as real in itself but as a social construct. The next chapter will analyze this notion and offer evidence that the conflict is, to a large degree, real.

Notes

1. Kwame Anthony Appiah, *Cosmopolitanism* (New York: Norton, 2006), xiv. See also Martha C. Nussbaum et al., *For Love of Country* edited by Joshua Cohen, (Boston, MA: Beacon Press, 1996), 6–11.
2. Jacques Derrida, *Rogues: Two Essays on Reason* (Stanford, CA: Stanford University Press, 2005).
3. Michel Foucault, *The Birth of the Clinic: An Archeology of Medical Perception* (New York: Vintage Books, 1975).
4. Richard Rorty, "The Contingency of a Liberal Community," in *Contingency, Irony, and Solidarity,* (New York: Cambridge University Press, 1989), 44–69.
5. Richard Rorty, *Philosophy and the Mirror of Nature* (Princeton, NJ: Princeton University Press, 1979) 174, 170.
6. Rorty, *Contingency,* 74.
7. Paul R. Gross and Norman Levitt, *Higher Superstition: The Academic Left and its Quarrels with Science* (Baltimore, MD: Johns Hopkins University Press, 1998).
8. John Caiazza, "Natural Right and the Re-Discovery of Design in Contemporary Cosmology," *The Political Science Reviewer* XXV (1996), 273–309.

2

Sources of the On-Going Conflict

The "War" as a Social Construct

Is the so-called "war" between religion and science not really a conflict but merely the result of a basic misunderstanding? There are reasons for thinking so. In contemporary terms, the war, or more properly, the conflict between religion and science can be termed a "social construct" which means that the conflict has no underlying reality except as each side represents a power structure. Such an interpretation is typically postmodern but has significant evidentiary support as can be seen in the politics of the original debate surrounding evolution. In Victorian times, evolutionists including Darwin himself were allied with the Whig Party and supported reform movements such as the abolition of the slave trade, while the anti-evolution party were largely Tories who supported the Church of England and the beliefs of upper-class England.[1] In the contemporary United States the controversy between evolutionists and creationists closely follows the "red state—blue state" fault line of American politics. Evolutionism is being advocated by educational elites who deem acceptance of evolution as the sign of a rational, educated human being; and creationism, in the form of intelligent design, is brought forth as a legal challenge by people who are usually middle-class and who represent the burgeoning power of the religious right.

A further indication that the war is a social construct comes from Stephen J. Gould who put the science/religion conflict in a historical context. He points out the influence of two nineteenth-century figures, John William Draper and Andrew Dickson White, whose writings became the basis of "the construction of the model of warfare between science and religion as a guiding theme of Western history."[2] For both writers, the enemy of rationalism was the Roman Catholic Church and theology generally—which was pictured as having destroyed the rationalism of the ancient Greeks by substituting blind faith, an institutionalized irra-tionalism that was overcome by the modern Enlightenment and the rise of modern science. In his enlightening historical account, Gould puts the writings of Draper and White directly in the context of nineteenth-century

Protestant anti-Catholicism. Readers can infer from his account that the war between religion and science came about not by any inevitable clash of principles and methodologies but as a matter of social attitudinizing, religious prejudice, and fear of waves of Catholic immigration then taking place in the United States.

So common are attempts to overcome the differences between science and religion that they have become the object of satire, as in the science fiction short story when a fictional Dean is launching into "version 3A" of his well tried talk; "There is no conflict between science and religion." Or in the Monty Python sketch in which a rationalist philosophy professor and an Anglican churchman go two out of three falls in a wrestling match. It has been said that the great nineteenth-century debate over evolution was merely a tempest in a Victorian teacup, that it was the result of mere misunderstanding, and that since science says nothing about those things which are truly at the heart of religious belief then there can ultimately be no conflict. Much intellectual effort has been put forth by recent philosophers, theologians, and scientists to overcome the perceived conflict that is not only between science and religion, but in a larger sense between science and liberal arts fields such as art, literature, and philosophy as bona fide sources of knowledge about the human condition. These irenically motivated thinkers argue in general that, once again, there is no conflict between science and the liberal arts; or if there is, it is merely the result of misunderstanding or hardened attitudes.

Despite such irenic efforts, the balance of evidence shows that the war between science and religion is, in some measure, a real conflict. War in general is a frightening thing whether in the real sense, involving guns, bombs, and troop movements, or in the analogical sense, in which widely differing principles and contradictory methodologies are used to understand the universe. Thus, many serious people, in considering the religion/science war, withdraw from it as if repelled and as if to say, "war is not the inescapable consequence of there being such prominent and important human activities as religion and science. Why can't religion and science just get along?" Yet despite the wishes of the well-intentioned or fainthearted, the actuality of the religion/science conflict cannot be denied.

Despite Gould's placement of the religion/science war in the context of nineteenth-century American history, science was allied with a rationalistic and liberal tendency that expressed itself in an aggressive public stance against not only the Catholic Church, but all of Christianity and, indeed, religion in general. The war between religion and science started not with the evolution controversy of Victorian times, but a century earlier in the days of the French Enlightenment in the eighteenth century. During this period philosophers such as LaMettrie argued that men were nothing more than organic machines existing in a universe of atoms swirling in the void; this was all perfectly understandable by purely mechanical principles and additions such as an immortal soul or a creator were not only unnecessary but positively hindered true understanding.

The best known attempt to directly unify the scientific and religious perspective is *The Phenomenon of Man* by the Jesuit paleontologist Teilhard de Chardin which posits an end to the evolutionary process in the form of the Second Coming of Jesus Christ.[3] Teilhard's mystical and teleological form of evolutionary theory became popular and remains influential but earned him harsh attacks from the defenders of both religious and evolutionary orthodoxy. Teilhard's religious superiors refused to give him permission to publish *The Phenomenon* (which was published only after his death) because of its Pelagian tone, while Darwinian evolutionists have either derided the book or attacked Teilhard personally.[4] The vehement reaction to Teilhard's attempt to reconcile revealed religion with evolutionary science shows just how difficult it is to unify religion with science in contemporary terms.

That the conflict continues seems undeniable from the recent wave of books, generally written from the perspective of evolutionary biology, which present reductive accounts of religion and the liberal arts; these range from the complex and sympathetic, as in E.O. Wilson's *Consilience*[5], to the aggressive and hubristic, as in Daniel Dennett's *Darwin's Dangerous Idea*. Dennett describes the theory of evolution by means of natural selection as a "universal acid" that will effectively destroy the rational basis of religious belief.[6] He gives this warning to believers: "If you insist on teaching your children falsehoods—that the earth is flat, that 'Man' is not a product of evolution by natural selection—then you must expect, at the very least, that those of us who have freedom of speech will feel free to describe your teachings as the spreading of falsehoods, and will attempt to demonstrate this to your children at the earliest opportunity."[7] Thus, if defenders of religion and cultural critics of science argue that there is no inherent conflict between religion and science, the defenders of a putatively scientific point of view obviously think otherwise. Influential promoters of the scientific point of view, such as Wilson and Dennett, not only think that there is an ongoing conflict between religion and science, they think their side is winning it.

Two Sensibilities: The Intuitive and the Rational

There are multiple reasons for the conflict between religion and science, and even if there is only an appearance of such a conflict or war, it remains to be explained. As described in the remainder of this chapter, there are three general sources of the conflict: religion and science are the result of two sensibilities, they present two more or less comprehensive worldviews, and they represent two different traditions in Western culture.

The vulgar caricature of the scientist in popular culture, as seen in science fiction stories, movies, and television, is that of an aloof, other-worldly figure of great intelligence. This person does not readily share relationships with other people and is ignorant of any connection, personal or professional, to the general society in which he lives. The portrayal can be benign as in the example of Dr. Zarkoff who assists spaceman Flash Gordon and his consort Dale Arden, or Mr. Spock, "science officer" of the spaceship Enterprise, who interestingly has no particular function on the spaceship; he doesn't do communications, engineer-ing, or navigation—just "science." Or the portrayal can be malicious as in various villains in the James Bond movies such as Dr. No, whose scientific expertise is used to destroy the efforts of legitimate governments and set up a technologically-based evil empire. More subtly, portrayals of scientists can be that of men of intellect whose powers are put to bad use because their vision is distorted, and here the iconic example is Dr. Frankenstein. Mary Wollstonecraft Shelley's famous novel, written at the beginning of the scientific revolution in the West, elucidated the options; scientific research must either be controlled by ethical constraints or it will give results which will terrify mankind.

Despite its melodramatic presentation, the Frankenstein story does reflect a scientific reality as we see, for instance, in the development of the atom bomb or in embryonic stem cell research. That is, regardless of ethical difficulties scientific research often proceeds simply because it is the next step in the development of scientific progress. When research impinges directly on ethical restraints, the research is usually carried out regardless of consequences or the likely immorality of the research itself. Scientific progress has its own impulsion we are told in such cases. Ethical objections are put out of court and declared irrelevant, often, as in the case of the fictional Dr. Frankenstein, with terrifying consequences. However arguable the Frankenstein characterization may be, it does point to a scientific sensibility that may be contrasted to a humanistic sensibility; therefore these kinds of characterizations are useful, to some degree, in evaluating the causes of the religion/science conflict.

Pascal, the seventeenth-century mathematical prodigy and religious writer, identified the two sensibilities as the "mathematical" and the "intuitive." According to him, the mathematical mind adheres to principles which are "tangible but far away from ordinary experience, so that, through lack of habit, it is difficult to look in their direction ... [b]ut in the intuitive mind the principles rest on common experience; and all eyes can see them."[8] We can easily translate this to infer a different kind of psychology or sensibility that propels science on one hand, humanistic studies including religion on the other, but which also applies to our present postmodern situation. Pascal's translator states: "The dispute between head and heart ... between an emergent science and a religion on the defensive, was not confined to Pascal's age. It continues, in a somewhat different form in our own, in which science is now on the defensive and a new religious movement as yet hardly emergent."[9]

The mathematical or scientific sensibility, utilizing abstraction and standardized methodology, deliberately excludes imaginative elements; it adhers to the objective as far as possible by means of close empirical observation and controlled experiment and always desires to express its conclusions in a definite form which can be taken to be precisely accurate if not final. It is skeptical, never taking first appearances at face value, as Copernicus saw the Sun as motionless in relation to a rotating Earth. It rarely if ever succumbs to emotion and distances itself from putative ties to individuals or social entities including family, church, or nation. It is ever ready to examine both sides in a dispute or alternative explanations and more readily concludes about general principles on its own than by receiving them from tradition or by instruction; Einstein began his theorizing by rejecting Newtonian mechanics and imagining himself riding on a light wave. The metaphysics of the scientific sensibility is restricted by its own approach that assumes but cannot prove that reality is measured in exact terms by the human intellect so that nothing is "left over" once the scientific truth of a matter has been determined.

The intuitive or religious sensibility relies on the immediate apprehension of reality that takes place within the subjective psychology of the individual; the experience occurs with such enormous psychological force that the reality of what is apparently designated is taken for granted. A world beyond the world of sense experience is manifested in such a way as appears indubitable. The experience is not necessarily or usually a single experience, although one may be enough to change a lifetime as it did for Arthur Koestler whose mystical experience while a prisoner during the Spanish Civil War led him to reject scientific materialism. It is usually a series of such experiences that may be of different degrees of intensity as in the case of philosopher Edith Stein who had left her family's Judaism and become an atheist but whose mystical experiences led her to became a nun and a saint. In general, such experiences lead to a conviction that while definite in itself, nonetheless requires explication that never reveals the whole content or tenor of the intuition. Thus, the experiences are described repeatedly and in different manners, giving rise to rich and varied forms of expression that present problems with the sharing or evaluating of such intuitional knowledge. Because of the indefiniteness of the insights gained through intuition, its conclusions are never really final, and the need for their continued reinterpretation forms the basis of an ongoing tradition of sustained commentary. The metaphysics of the intuitional sensibility is indefinite but assumes that reality is not fully compassed by human knowledge and never can be, that it lies beyond the specifically material.

The separation of a scientific from a religious sensibility may seem in Pascal's writings to be merely a quick comment or *apercu,* and so it worth pointing out that such a distinction is the basis of Kant's entire philosophy. Kant, whose writings remain a dominant influence on philosophy, was sensitive to the success of modern science as exemplified in the new mechanical view of the world and so segregated modern science as a special form of knowledge different from that which was the basis of philosophy and ethics. Thus, Kant specified a "transcendental aesthetic" which described the categories of time and space as the basic underlying elements of Newton's mechanical system. In using the term "aesthetic," Kant meant to point out that modern science was based on a worldview which could not be irrefutably attached to our knowledge of reality. He termed this ultimate reality "noumenal," since it was a realm which remained beyond penetration by the human intellect. Ethics and religion required, for Kant, a secondary type of knowledge, one that was practical, a matter of judgment, and not yielding the type of certain, mathematical laws. As for religion, Kant dealt with it purely "within the limits of reason alone," not allowing for any revealed component. Nonetheless, he effectively attempted to base Christian ethics on a rational basis by deriving the principle of the *categorical imperative*: "Act only on a maxim by which you can will that it, at the same time, become a general law," From this he derived a *practical imperative*: "Act so as to treat [all persons], in your own person as well as in that of anyone else, always as an end, never merely as a means."[10] This, however, is not the place for a disquisition on Kantian ethics since it is enough to recognize that Kant's entire philosophy was effectively based on the distinction between what Pascal had termed the contrasting "sensibilities" of mathematics and intuition.

Understanding the conflict between science and religion as a conflict between sensibilities does not mean that there are two different personality types, but rather that there are two aspects of human personality in general. To paraphrase Solzhenitsyn, the line between the mathematical and the intuitive sensibility runs not between social classes or different professions but through each human heart. What has happened in modern times, however, is that as society and social roles have become increasingly differentiated so the scientist and the humanist have their own separate professional functions, academic departments, and social distinctions. Despite the differences of sensibilities, it is notable how they often overlap within one personality so that we have scientists who are true mystics who rely on their imaginations to produce theories. Thus, Faraday, inventing force fields as a means of unifying all natural phenomena, and Kepler, devising geometrically defined physical laws as an expression of divine harmony, produced valid scientific theories which are founded on their personal intuitions. We also have examples of theologians so rationalistic in their approach as to be accused of having excluded emotion from religious belief and reduced the transcendent aspects of revelation to a closed intellectual system. These accusations were made against St. Thomas Aquinas whose *Summa Theologica* resembles a comprehensive textbook of analytic philosophy and Reformer John Calvin whose *Institutes of the Christian Religion* resembles a tightly drawn treatise on civil law.

Two Comprehensive Worldviews

The concept of worldviews has become commonly accepted as a means of explaining how different people can come to diametrically opposed opinions on particular issues because the issue upon which disagreement rests is part of a system of opinions about issues that are interrelated according to a roughly definable set of general opinions or first principles. Religion and science as worldviews can be contrasted in this manner.

Religion and science provide widely contrasting yet eerily similar general explanations for the universe as we describe and experience it. The religious, that is biblically grounded explanation, (I will deal in the thirteenth chapter with the differences among religious traditions) of the universe is based on the creation story in *Genesis* and on the consistently moralizing approach taken in explaining not only historical events in the life of ancient Israel, but the details of the physical environment and of species as they were known in that ancient time. Things happen or things exist as they do basically because the Lord intended that they happen or exist in that way. All things are ordered to the plans of the Almighty: from the stars in the sky, the rains that come from the sky, the rivers and oceans that God set apart from the dry land, the Sun in day or the moon at night set in their appointed place to give light to mankind, and the fish in the sea or the animals on land that the Lord gave for mankind's use. And those items which have no obvious practical use, such as lilies of the field or sparrows of the air, are useful as the basis of parables to show the providence of God and His love for his fleshly creatures. The iconic illustration is provided by the English artist and mystic William Blake and shows a bearded God bending over to observe his creation, as he lays out the foundations of the world with his finger. (Now, of course, we would picture Him as a bearded old man sitting at a computer console, tapping into the "exec" files that will determine the laws of the universe.[11])

The scientific explanation of the universe is objective in intent, with that kind of dispassion and distance from moral and emotional connections, which has come to characterize modern science. The scientific explanation is based on the discovery of laws of nature which determine the behavior of material entities in such a way that can be predicted with very close and useful accuracy. The laws themselves are built up from empirical observations and measurements from which patterns are discerned and are then put into mathematical form. Such was the procedure of Galileo and Newton, and then all the scientists following in a train which leads through E.O. Wilson and Steven Weinberg up to this day. Ultimately, science seeks to explain everything in the universe according to a fully integrated set of empirically based but mathematically abstract scientific laws. The difference with the religious mode of explanation is evident in that moral causes, such as punishment for sins or providential provisions for human existence, are never considered and that final causes and miraculous interventions are excluded as a matter of explicit method.

The term "scientific method" is well known and religion, it is true, does not have a "method" as such, for one thing, because religions are so diverse and varied being that they are tied to cultures or universal associations. But religion, it can be observed, generally depends on interior beliefs and experiences which do not lead easily to the kinds of generalizations that science makes and which are, in fact, beyond the explicit range of logical or empirical proof. This does not leave religion without a method of establishing doctrines and being able to promote its belief systems to skeptical people who are possible adherents, however. Typically, there is a text, a holy book, believed to be revealed by God or through divine agency which provides a guide, or *proof-texts,* for discovering and proving particular doctrines. Socially, the doctrines are made part of a cult which involves a group of believers in a membership that is exclusive and which also provides a means of guardianship of the purity of doctrine. Apostasy and heresy are typically condemned as the worst of sins, and people who participate in such are shunned or more severely punished if the religious authority has influence in the governmental structure of society. The truths of religions are typically carried on by means of a tradition; they are written down, passed on orally, or conveyed through the special practices of religious worship, e.g., reciting the Nicene Creed or recounting the Passover story.

Science's method is twofold, depending on both an empirical aspect, which concentrates on the material entities and the observable processes of the physical universe, and also on an abstract space in which mathematical laws and general scientific theories exist. That science's methodology has a double aspect involving both the empirical and the abstract is a point often lost among English and American philosophers who assume modern science to be the example above all other examples of empirical methodology, an assumption that underlies the tendency toward materialism and naturalism. While it is patently true that science is devoted to the discovery of physical phenomena, it also involves the intellectual process of generalization. For instance, it recognizes that certain individual birds make up a separate species but then goes beyond to discover increasingly abstract ranges of generalization and law—from species to genus, phyla, and kingdom moving up the abstract ladder of classificatory categories-and then uncovers relations among the categories, so that species make up not just a static pattern but are in a dynamic relation to one another. Scientific method thus depends not only on discovering the physical truth through experiment and observation but subsequently on making inductive generalizations, matching the generalizations against the current theories, and redoing the experiments or altering the theories as needed—a process of mutual correction that is the core of scientific development. The general methodology of science depends heavily on the transmission of its discovered truths as much as religion since without persistent institutional memory scientific discoveries would be worthless. While there is no reliance on tradition as such in modern science, its success depends absolutely on the continuing transmission of information through university and college courses and textbooks on the various fields of science, in journals and books, on-line, as well as in archives held by professional associations and independent research organizations.

What gives the religious and scientific modes of explanation their eerie similarity, and also the basis for their conflict, are the two following factors. First, both the religious and scientific worldviews attempt to explain the universe in detail and not merely in general terms so that, for example, the fact that a person has acquired an sexually transmitted disease is explained religiously, the person violated norms of sexual morality, or scientifically, the passage of a viral infection by means of intimate contact. Second, both portrayals are *historical*, each giving an account of the universe extending from its very beginnings—as when God said "Let there be light" or from the time of the Big Bang to the projected end-times as described in the *Book of Revelations* of St. John which describes the Second Coming of Jesus Christ or as speculated on by contemporary astronomers in terms of "heat death" or an ultimately explosive "big crunch"; these are the two likely ways in which the physical universe comes to its inevitable end.

The sharp differences between the general methodologies of religion and science reflect what above was referred to as their opposing metaphysical assumptions. Thus, for religion, the ultimate truth is reflective of a reality that escapes human intelligence at certain vital points. For example, God's plan for us and his direct involvements in human existence must be revealed to the human race, and as a practical matter this requires that revealed truth must be maintained through some type of direct social control. (This point has been made by critics of religion including Russell and Popper.) For science, the implied metaphysics is that reality is if not perfectly than enough so within the powers of human intelligence that whatever reality is not attainable by the human mind may not count for much and is probably irrelevant.

Two differing ethics are proposed by the religious and scientific worldviews; it is among the functions of a worldview that it offers an ethic—not in the formal sense, necessarily, of a defined system but in the general sense of providing a guide to practical living. Religious ethics are often what is called "command ethics," that is, divinely given rules for behavior to which are attached sanctions imposed either by religious society or in the afterlife by the gods or by God. There are many examples of divine commands presented as ethical guides, and Plato in the *Euthyphro* deals critically with an ancient Greek version. The Ten Commandments presented by Moses as divine prescriptions are found in the book of Exodus in the Hebrew Bible and address such things as swearing evidence in court, theft, and adultery. But the commands regarding behavior towards other people are preceded, in the Hebrew version, by commandments about how people are to behave toward God—rejecting idolatry, for example, the implicit idea being that from God flow not only the commandments but that from the holiness of his nature flows the reality of life as it is meant to be lived.

There are no ethical commands emanating from the scientific worldview as there is no putative agent to give them. However, since science is a social enterprise set in a special manner within society and has an elevated role within it, the scientific worldview implies certain general rules about what constitutes good human behavior. Scientific research, which is very expensive to maintain, is promoted as the means of improving the quality of human life as well as providing the best means of understanding the physical universe. Science's situation in society implies a point of view that the highest good is to provide practical relief of life's tragedies and difficulties such as poverty, illness, ignorance, and death. But for science to continue and produce its remarkable results, both practical and theoretical, certain values are required of the scientists themselves—such as honesty in reporting research results, diligence in carrying out experiments, not plagiarizing the work of other scientists, and giving due credit to research assistants and predecessors.

Conflicts between the ethical consequences of the two worldviews have become apparent in contemporary society involving issues of sexuality including abortion and the use of condoms. In these issues, the practicality of scientific ethics comes in conflict with the imputedly divine source of religious ethics; from the point of view of scientific ethics abortion and condom use are effective means to rational ends, i.e., preventing unwanted births during a time when overpopulation is a concern and preventing the spread of AIDS, still an inevitably fatal infection. If ethics is based on what is defined as the good for human beings, living one's life fully in a material manner is the good in scientific terms while living in accordance with the divine law is the good in religious terms. But this difference, in turn, implies a difference as to what human nature is in essence, either a complex set of naturally evolved systems or an image of the divine reality.

Two Differential Traditions in the West

The ongoing conflict between religion and science in the West in large part depends upon historical tradition—tradition rooted intellectually in the development of modern philosophy. The standard account of the history of Western philosophy posits a great separation at the end of the medieval period that marks the beginning of the modern period in Western thought.[12] At the end of the medieval period comes the disintegration of the holistic account of the universe which combined Aristotelian metaphysics with biblical revelation—an intellectual feat accomplished by representatives of Judaism, Islam, and Christianity. Theologians such as Maimonides, Averroes, and Aquinas had been struck by the fact that Aristotle and other ancient thinkers had developed a notion of God characterized as one, immaterial, eternal, and unchanging and who was responsible for the providential running of the universe to insure the existence and comfort of mankind. The ancient Greek thinkers had created this *natural theology* completely without benefit of revelation, i.e., knowledge of sacred scripture. Aristotle had extended his metaphysics into a comprehensive philosophy of nature, for he was an experimenter and researcher of animal bodies and is considered the first biologist (as such, his name is carved on the lintels of the buildings at MIT along with those of Newton, Galileo, and Darwin). Although knowledgeable about biology, Aristotle's account of the basic physical universe is rejected by modern science, and his account of the motion of the sun and planets was geocentric, not heliocentric. However, it was the integrated aspect of medieval thinking, combining as it did philosophy, science, metaphysics, and the truths of the Bible, that made the intellectuals of Galileo's time resistant to abandoning Aristotelian account of physical motion and the geocentrism, which Aristotle had taken from the ancient astronomer Ptolemy. Galileo, however, abandoned both and argued effectively against Aristotle's theory of motion and his Ptolemaic astronomy.

The breaking apart of the medieval synthesis came about because of the rise of modern science, and the foremost intellectual expression of the new, somewhat discordant outlook came from the writings of Bacon, Galileo, and Hobbes but most especially Descartes. It was Descartes who, in his short book *Meditations,* separated reality into the mental arena of spirit and thought, on one hand, and the physical arena of material reality, on the other. The former included mathematics, philosophy, the spiritual aspect of the human soul, and the existence of God for which he offered dialectical proof, i.e., without benefit of external evidence from the senses or experience. Physical reality was a literal afterthought for Descartes since he asserted that only because God insured its reality, could we prove dialectically that the external world, detected by our senses, existed. The dependence of the truth about physical reality on mental reality is an aspect of Descartes' thought that is not usually taken into account enough by Descartes' philosophic critics and possibly not by Descartes himself. However, it is true that by dint of his reasoning Descartes laid the foundation for the belief that study of the physical world in a detailed manner was possible and that truths garnered by scientific study were more certain than those garnered from dialectical philosophy. Later generations would reject Descartes' comparison, asserting that spiritual, metaphysical knowledge was only subjective and thereby untrustworthy and that only the discovery of scientific laws gave mankind access to certain truth. In whatever manner Descartes' philosophy is evaluated, it is true that after him Western philosophy ran on a double track, rationalistic on one hand, and empiricist on the other.[13]

The ultimate effect of the breakup of the medieval worldview on the development of the conflict between religion and science becomes apparent—for the putative objects of religious belief are relegated to the mental realm, now understood to be subjective, while reality is increasingly assumed to be revealed by empirical science in terms of its mathematical laws. Rational experiment and well defined logical inference leading to general laws is the apparent essence of science and the objects discovered by scientific research; phenomena like the circulation of the blood, things like the moons of Jupiter, or scientific laws such as universal gravitation and the germ theory of disease are presumed to be the real stuff of which the universe is made. On the other hand, the belief system of religion becomes a matter of subjective belief because immaterial entities such as the soul, God, or salvation, which are the objects of traditional religious belief, are in principle undiscoverable by scientific methodology. More generally, it is said that values are not logically entailed by physical facts; in the formulation of Hume, much repeated as an essential part of the philosophy and methodology of social science, one cannot derive an "ought" from an "is." Once accepted on its own terms, the scientific worldview became the total environment of human existence, and any sense of purpose or providence, much less that the physical universe was the creation of God, was lost or, more precisely, explicitly rejected without any temptation to nostalgia or regret. Necessarily, values are thought to be merely subjective, opinions, or sentiments imputed to the physical universe without intellectual warrant. The belief that science will not logically support a valuative view of the universe implies that the immaterial objects of religious belief lose veridicality. This belief has become the major cause of the dissociation of the religious and scientific points of view. Leo Strauss commented that until that breach is overcome, western humanity is destined to have a divided view of self and the universe and will be doomed to the kind of dualism first envisioned by Descartes.[14]

The idea that the conflict between science and religion has an historical basis that starts from the time of Galileo and Descartes, which involves the breakup of the medieval worldview and the development of modern science, constitutes what can be termed "the standard view." However, the present author has argued that the historical basis of the religion/science conflict can be pushed back into the history of the West—into the third century A.D. by reference to what some have termed the double tradition of Western history; the template of Athens versus Jerusalem.[15] In this portrayal, the identity of the West is constituted by a tension between Athens, representing reason and philosophy, and Jerusalem, representing faith and religious belief. The notion that there are irreconcilable differences between the intellectual and secular traditions of the West and the faithful pursuit of the holy begins with a complaint by the theologian Tertullian in the fourth century. He asked rhetorically, "What has Athens to do with Jerusalem?" making a point that has been repeated throughout the history of the West to the present day when believers state that the knowledge procured from the Bible or the Koran, being enough for personal salvation, trivializes all forms of human knowledge. In the fundamentalist believer's formulation, Athens is rejected in favor of Jerusalem, but of course the opposite tack is often taken by scientific advocates as in the reductive explanations of religious belief common today among the intellectual classes of the formerly Christian West.

Less oppositionally, authors such as Leo Strauss and Jeffrey Hart have argued that the chief characteristic of Western civilization is that it contains both the elements of Athens and Jerusalem in tension.[16] This understanding of the nature of the West implies that things work best when neither secular philosophy nor religious faith are dominate. The medieval theologians who attempted to combine Aristotelian philosophy with biblical revelation or St. Augustine who wrote Christian theology from a Platonist perspective are representative of the attempt of integration of the intellectual/secular with the faithful/religious points of view. It remains to point out that modern science can be seen as an extension of the Athenian tradition in the West—as a secular and intellectual attempt to understand the universe and control the physical environment. The current debates between religion and science indicate, however, that an intellectual integration with biblical knowledge is not so easily done with modern science as with the classical and modern traditions of philosophy. Philosophy, unless explicitly materialistic, has room for the ineffable, if not for the one God or an explicit religious creed, making it possible to explain religious faith from a Platonic, Aristotelian, and also from a Rationalist (Descartes), Idealistic (Hegel), Phenomenological (Lonergan), Analytic (Anscomb) and Existential (Tillich) point of view. Modern science, however, since its method rejects miraculous and teleological explanation on principle is impossible to integrate with a biblical account of the universe—that is when science is taken in its materialistic mode.

Taking the history of the West going back to the third century as the basis for the current conflict between religion and science implies that modern empirical science continues the tradition of Athens. However, the Athenian tradition up until the arrival of modern science could be integrated with theology if the theologian was intelligent, creative, and learned. Modern science because of its inherently naturalistic bias puts up a barrier between the principles of religion and those of science.

*   *   *

In summary, the question of whether there exists an inherent conflict between the traditional religions of the West and modern science seems to have been answered; there are objective grounds for the conflict and the perceived conflict, therefore, has a basis in reality. As has been presented here, the conflict between religion and science rests on their being based on differing sensibilities: the intuitive and the mathematical; on their offering competing worldviews that attempt to explain the universe in detail based on different principles; on differential traditions in the West whether they emanate from the time of the break-up of the medieval worldview or the tension between Athens and Jerusalem originating in the third century A.D.

It may be, as some of the most serious and learned authors who have considered it have concluded, the conflict between religion and science has no objective basis in fact; however, this assertion rests on the possibility that there is an overarching intellectual system containing the principles of both religion and science. I have argued that the difference is objective, and if so a unification or reintegration of knowledge as existed in the Middle Ages is impossible—that is if a materialistic and deterministic view of science is held. But science itself has taken a postmodern turn and has now passed from its formerly modern phase into a new phase, possibly even an end phase, which science writer John Horgan terms "ironic science."[17] The new phase means that science is no longer properly called "modern" but can be termed "postmodern," and so we look in the following ten chapters at the shifting ground of the religion/science conflict based on new theories and ideas from within science itself. Then, in the final three chapters, we will conclude by looking at what a reintegration of religion and science might look like in the postmodern world.

Notes

1. Gertrude Himmelfarb, *Darwin and the Darwinian Revolution* (New York: Anchor Books, 1962), 255–287.
2. Stephen J. Gould, *Rocks of Ages: Science and Religion in the Fullness of Life* (New York: Ballantine, 1999), 117.
3. Pierre Teilhard de Chardin, *The Phenomenon of Man* (New York: Harper and Row, 1959).
4. Peter Medawar's attack on *Phenomenon* is a remarkably hostile review (www.umich.edu./~crshalizi) while Gould accused Teilhard of complicity in the Piltdown hoax. See Gould, *The Panda's Thumb* (New York: Norton, 1980), 100–124.
5. Edward O. Wilson, *Consilience* (New York: Vintage, 1998).
6. Daniel Dennett, *Darwin's Dangerous Idea* (New York: Simon and Schuster, 1995), 521.
7. Ibid., 519.
8. Etienne Pascal, *The Pensees,* translated by J. M. Cohen (Baltimore, MD: Penguin, 1961), 33, 34.
9. Ibid., 10.
10. Immanuel Kant, "The Metaphysical Foundation of Morals," in *The Philosophy of Kant,* edited and translated by Carl J. Friedrich (New York; Modern Library, 1949), 170,178. Compare *Luke* 10:27, *Leviticus* 19:18.
11. John Caiazza, *Can Religious Believers Accept Evolution* (Huntington, NY: Troitsa Books, 2000), 132.
12. W. T. Jones, *A History of Western Philosophy; Vol III, Hobbes to Hume* (New York: Harcourt World, 1969), 1.
13. Barbour, *Religion and Science.* 12, 13.
14. Leo Strauss, *Natural Right and History* (Chicago: University of Chicago Press, 1953), 8, 9.
15. John Caiazza, "The Athens/Jerusalem Template and the Techno-secularism Thesis," *Zygon* 51, 2 (June 2006), 235–248.
16. Jeffrey Hart, "What is the 'West'?" *Modern Age* 47, 4 (Fall 2004), 362–366.
17. John Horgan, *The End Of Science* (Reading, MA.: Addison-Wesley, 1996).

Part II

Shifting Grounds of the "War" between Religion and Science

3

The Fog of War

The war between religion and science began in the nineteenth century. It did not begin in the seventeenth century with the rise of the "new physics," when mathematical descriptions replaced qualitative descriptions of motion, nor does it exist in the same manner today as in the nineteenth century since the rise of the "new" new physics of relativity and quantum. The grounds for the persistence of this war, despite doubts about its ultimate reality, were presented in the previous chapter. While the "war" between religion and science is a metaphor, it is a war in the sense that certain social authorities, received traditions, ancient philosophies, and mere bigotry are said to be aligned in a systematic way against modern science, which includes determinism, materialism, and atheism—the inevitable consequences of hubristic science. This double stereotype continues to exist in many people's minds, but if the apparent war was not the result of implacably aligned intellectual forces on either side, then the war is more of a perceptual effect or, as the postmodernists put it, a social construct that does not reflect basic reality. The constructors of this war were nineteenth-century intellectuals including the first President of Cornell University, Andrew White, whos influential book, *A History of the Warfare of Science with Theology in Christendom,* presented a progressive view of the advance of science overcoming challenges posed by orthodoxy. As a "Whig history," it characterized the path of science as a series of triumphs of "science" to the detriment of "theology." To quote White:

> In all modern history, interference with science in the supposed interest of religion, no matter how conscientious such interference may been, has resulted in the direst evil, both to religion and to science ... on the other hand, all untrammeled scientific investigation, no matter how dangerous to religion some of its stages may have seemed for the time to be, has invariably resulted in the highest good both of religion and of science.[1]

Stephen J. Gould pointed out that, like himself, White was concerned with true religion, which he did not wish to see injured, and which both White and Gould agree can never be overcome by science. For them, true religion is to be distinguished from "theology" which makes declarations about such subjects as the creation of the universe, the origin of life forms, and certain "miraculous" events that science has discredited. Once theology is distinguished from true religion, the so-called war between religion and science can be called off. This is Gould making his point while letting his mask of avuncular friendliness slip:

> Many religious intellectuals have always been happy to cede inappropriate territory to the legitimate domain of science, but others, particularly in positions of leadership, chose not to yield an inch, and then played the old hand of dichotomy to brand the developing magisterium of science as a sinister bunch of usurpers under the devil's command – hence the actual and frequent warfare of science, not with religion in the full sense but with particular embodiments better characterized as dogmatic theology, and contrary to most people's concept of religion ...[2]

The view that White and Gould hold of true religion is rather toothless, not much more than a form of moral uplift and encouragement to social reform. As one of Gould's critics pointed out, his notion of religion did not include the belief in God. Certainly it is true that in their version there should be no war because science has already won all the important points, and there is nothing left for the proponents of revealed religion to do but, like Caesar's conquered Gauls, accept the reality of foreign imposed rule and lay down their intellectual arms. In any case, Gould concedes that in an historical sense there has been, and continues to be, a "war" between science and religion and the remaining question is not whether it exists but what manner of war it is.

The war can be analyzed as taking place in two arenas at once: on the intellectual level, where statements and assertions by religious authorities directly contradict those of scientific authorities and vice versa, and also on the social level, where the forces institutional religion and secular society have contended with one another. On the intellectual level, the war should not exist because there are coherent and plausible philosophies, theologies, and intellectual structures that can encompass both, and, therefore, it is largely a war of competing social identities. The fact is not often enough appreciated that the early giants of science, such as Kepler, Newton, and Galileo, were religious believers who readily accepted the basic doctrines of the Christian religion. It seems that for them there was no conflict intellectually between revealed religion and the new mechanical physics, which was opening the door to an exact description of the physical universe in terms of mathematical laws and close empirical research. It would be contemptuously condescending and ahistorical to impute ignorance to these men as if to say they lacked the critical reasoning to see or perceive the inherent intellectual conflict or decided to repress it because of lack of intellectual courage. Newton was so convicted of his Puritan beliefs that he wrote theology more often than he wrote mechanics, and Galileo entered fully into the theological controversy about the proper use of sacred scripture in scientific research (see "Galileo's Enduring Career," chapter 6).

For seventeenth-century scientists such as Galileo and Newton, who founded the new mechanics and astronomy, and for later scientists including Faraday and Maxwell, who founded the modern theory of light and electromagnetism, there was no intellectual conflict between revealed religious doctrines and the facts and theories of modern science since, for them, the God revealed in the Bible and Christian tradition was God the Father, the Creator. For these early giants, doing science was a way of discovering the immutable laws on which God had made the universe, a process of discovery that was a form of praise, therefore, of the Creator. That it was the job of science to discover immutable laws of physical creation is reflected in Einstein's remark that he was trying to discover if God had any choice in the way he had created the universe. Einstein, however, was not a believer in revealed religion having rejected *the Torah* and teaching of Moses in his youth as irrational contrasted to the truths revealed by science. Einstein along with Darwin, who also rejected biblical faith, have become the symbolic representatives of modern science's apparent rejection of revealed religion. However, it was not so in the beginning and professional agnosticism did not become the badge of scientific identity until late in the nineteenth century. Both Einstein and Darwin were influenced by family attitudes in their rejection of revealed religion. Einstein's family, while Jewish, was not observant, and Darwin's father and brother were both nonbelievers. It is not difficult to believe, also, that both Darwin and Einstein were heavily influenced by the ambient culture of nineteenth-century Europe. It elevated scientific rationalism above religion, which was seen as a phase—as it were of humanity's childhood that the human race had outgrown—according to the influential account of August Comte, the French inventor of sociology and positivist philosopher of science.

What was typical of nineteenth-century science in its cultural context was its determinism–for science had compassed seemingly the entire world not only abstractly, with advancing theories, but practically, in terms of industrial applications. While Newton had written his *Philosophiae Naturalis Principia Mathematica* as a philosophy of nature just as specified in the title, later generations extended and refined his intellectual edifice to make it more true to scientific fact but void of its philosophical aspect. Pierre Simon de Laplace, one of Newton's successors, early in the nineteenth century gave the new vision the name *The System of the World*, and corrected Newton's synthesis of astronomy, mathematics, and mechanics by, among other things, explaining why the planets did not exactly follow an elliptical orbit as predicted by Kepler's Laws. Laplace also made explicit the thesis of determinism, saying that if an observer could have knowledge of the position and velocity of every particle in the universe, that the state of the universe at any later time could be predicted.

> Let us imagine an Intelligence who would know at a given instant of time all forces acting in nature and the position of all things of which the world consists; let us assume further that this Intelligence would be capable of subjecting all these data to mathematical analysis. Then it could derive a result which would embrace in one and the same formula the motions of the largest bodies in the universe and of the slightest atoms. Nothing would be uncertain for this Intelligence. The past and the future would be present to its eyes.[3]

The subtext of Laplace's thought is that physics can apply the deterministic equations of Newton's differential calculus to individual atoms, which are assumed to be extremely small, slightly elastic particles, whose nature is not essentially different from that of common matter that makes up trees, rocks, flesh, butter, dirt, houses, and baseballs. In this way it was theoretically possible for scientists to know all things in the universe's past, present, and future. Thus Laplace stated by encompassing all the mathematical laws that determined the paths of atoms and all larger bodies into one comprehensive formula–the ultimate vision of deterministic science. This too was an idea implicit in Newton's *Principia,* but what Laplace left undetermined was the nature of that "Intelligence" whose name he capitalized. Was the Intelligence the infinite mind of God? The answer was "no." For Laplace, in response to a question from Napoleon, stated in reference to God that he "had no need of that hypothesis," thus reducing the notion of God from creator to an extraneous extension of physical theory. Newton had claimed that the intervention of God was necessary to fix abnormalities in the orbits of the planets, an effect called "perturbation." which Laplace neatly explained as the mutual attraction of the planets–so that Jupiter, for example, which is the largest of the planets, pulled Mars slightly from its elliptical orbit. Laplacean determinism in physical theory eliminated the agency of the revealed God completely, along with human free will, and the intervention of any other nonphysical agents in the course of the ongoing actions of the universe. This was, however, a methodological assumption which was also a reflection of the overestimation of the power of the human intellect and the desire for complete mastery typical of scientific philosophers.

Throughout the nineteenth century the hubris of scientific philosophers and many scientists made it seem possible that the precise contours of a self-dependent universe were being discovered, and, as the scientific account was being filled out in detail, it indeed competed with the account given in the Bible and would displace it. Like sixteenth-century nobles taking over monastery lands, scientists in the heady days of nineteenth-century colonial expansion were taking over the explanatory territory formerly occupied by revealed truth. But the nineteenth-century deterministic view, which characterized the physical universe by machine-like properties amenable to science's virtually infinite ability to understand physical reality, was to undergo radical revision in the twentieth century. The development of quantum physics and relativity theory rendered a picture of the universe which was indeterminate on the micro level, i.e., in its atomic structure and chaotic on the macro level, as a flood of possible geometries of space-time, alternate universes, and the Big Bang. Choice of theoretical options became the main feature of how physicists now understand the physical universe.

Theoretical physicists in the twentieth century, reading Laplace's account of an "Intelligence," would be skeptical of the Intelligence's extraordinary, indeed, infinite ability to know the position and motion of individual atoms—a problem of real importance in twentieth-century atomic physics. Unless Laplace's Intelligence was in fact God, according to the Heisenberg uncertainty principle, it could not know both the position and momentum (a property which includes velocity) with absolute precision. As it was defined in mathematical terms by Heisenberg, the more data you had about position, the less you had about momentum and vice versa. This conundrum was only one aspect, however, of the contradictions to Newtonian mechanics and common sense that emerged from research conducted from the very end of the nineteenth century until present day about atomic behavior. Position and momentum were *complementary* properties according to the new quantum physics, which notoriously placed probability at the heart of physical reality, i.e., within the internal structure of the atom itself. The elemental and homogenous atoms of Newton and Laplace were dissolved by a series of intricate laboratory experiments and new theories into a conceptual "zoo" of particles and forces so complex and intertwined that their relationships and behavior defy deterministic understanding. Modern, classical physics has now ended and a new postmodern physics has succeeded it. The problem of the "observer," which classical physics ignored, is central to modern physics, which must take account of the observer in its equations, much like the social sciences in fact.

In a famous experiment done at a Westinghouse factory in the 1940's that was meant to determine if giving workers such amenities as snacks, rest periods, and bathroom breaks would increase productivity, the researchers discovered that it made no difference. There was an increase in productivity and an improvement in attitude both for the group that was offered these amenities and the control group, which was denied them. The increases in productivity came about not because of the availability of amenities but because researchers had spent time advising and questioning the men about their reactions. The workers then became convinced that someone was interested enough in them to analyze their work and ask their opinions which was the factor that motivated them to a more productive attitude. In short, it was the presence of the sociological observer, not the snack period, which determined the outcome of the experiment. This so-called "Westinghouse Effect" is present in the physical sciences as well. The presence of the observer in the experimental setting must be accounted for in quantum physics and also in relativity.

Relativity theory is not concerned about the observer who must detect an atom but rather the observer who must define an entire astronomical universe. Twentieth-century physics dismissed Newton's assertions of absolutes in time and space, which meant that, lacking a universal frame of reference by which to describe physical experiments, definition devolved to the single observer who had to utilize his own position as the point from which to describe the universe. Oddly, this meant that in relativistic terms, the Ptolemaic astronomer has as much a "right" to his account of the relative motion of the Sun and planets as the Copernican astronomer. This is not to say that all is subjectivity and chaos however, for Einstein had declared that the speed of light was constant in all frames of reference. Also, he maintained, that there were equivalences that could be systematically made between frames of reference, i.e., between the space-time systems of different observers, for example, the equivalence between mass and energy. Even in quantum mechanics, while the mathematics and perhaps the ontology are probabilistic, the scientific relationships are precise; otherwise, they could not be used for prediction—not to mention in the design of computer chips. Relativity might make the controversy between the Ptolemeans and the Copernicans moot, according to the late astronomer royal, Fred Hoyle, but only because one system could be transformed into the other with mathematical precision. However, the very equations of basic physical theory must now encapsulate the existence of the observer—as quantum phenomena are said to "collapse" upon observation and relativistic equations describing length, mass, and velocity must take into account the speed of the observer in reference to the speed of light.

At the beginning of the twenty-first century, the physicist's universe is less like a machine than an organism, i.e., an entity that has a beginning in time, which has undergone expansion, growth, and increase in complexity; it is currently in middle age and will eventually die. Furthermore, all the physical events in this new universe reflect not the necessity and optimism of the nineteenth-century picture of the universe but rather the anxieties and contingencies of the twentieth. The place of the individual physical observer is no longer an assured one in which the actions of the universe pass before him, as in a play according to a script, which can be observed from out beyond the proscenium by the scientist. In the postmodern era, the physical observer becomes part of a Pirandello play, interacting with the characters and unsure of what response to make, a condition that now extends beyond even theoretical considerations of quantum observation or relativistic time constancy as the cultural, ethical, political, and religious contexts have entered the heart of the scientific enterprise.

With the role of the observer and indeterminism now a part of science's view of the physical universe, contingency and anxiety become central aspects of this same universe, and thus the place of religion, vis à vis the physical universe, necessarily undergoes alteration. If scientific determinism and materialism in the nineteenth century can be associated with rationalism and social progress, then indeterminism and the role of the observer can be connected to the free-floating anxiety characteristic of the twentieth and twenty-first centuries. This is a psychological and cultural affect brought about, in large part, by scientific improvements of the weapons of war—from machine guns to nuclear bombs. In the context of contemporary culture, while battles between religion and science still take place, the lines of conflict and the ground on which they stand have significantly changed.

Battles between evolutionists and creationists take place with evolutionists claiming the complete intellectual victory of what may termed genetic imperialism (as evolution has been extended from the forms of organisms to explain their behavior), while the creationists promote a scientific form of creationism that perceives the handiwork of God in the passages from one evolutionary stage to another. Within physics, attempts to encompass relativity and quantum theory into a unified theory of physical reality (still the ultimate aim of science) have ascended into regions of abstract mathematics. These regions talk of strings and eleven dimensions, of multiple universes and a single combined force; these theories, however mathematically interesting, have less and less to do with actual experiment and observation and are becoming so speculative as to resemble not even metaphysics but science–fiction. One actual research result that is claimed will help validate one version of an ultimate theory of everything is *proton decay*. However, since protons are the basic part of atoms (on one level), the fact of their decay heralds the end of the physical universe. Extensive experiments involving millions of gallons of purified water hidden underground in salt mines have been set up to discover if protons actually decay (it does not happen very often if it happens at all). But like General Lee and his army, physicists must now envision as the end point of their professional activity—the destruction of that thing they are most dedicated to, the material universe.

As a result of these developments, science has retreated to some degree in the war with religion. As seen with the study of physics, namely, the physical universe can no longer be understood as a self-subsistent entity. Some physicists as a result have begun to engage in "God-talk," speculating about the existence of God and the possibility of detecting purpose and design in the universe (see "Why the Physicists Speak of God," chapter 7). With a beginning and an end, our universe as described by postmodern physics is no more permanent than any smaller part of it—whether galaxies, planets, or soap bubbles—all of which have their own course of birth, growth, and death. But the impermanence of the material universe has been a constant subject of the revealed religions, for the Bible speaks both of its beginning in Genesis and its end in Apocalypse. Thus, it would seem that at this point in cultural time theologians and scientists would have something to say to each other on the topic of the fate of the universe. As for the free-form psychological anxiety and sense of cultural disintegration, the primary strategy of religion is to contrast the eternity and love of God with the impermanence and coldness of the physical universe. So on the psychological level as well it would seem to be time for social scientists and religious believers to consult with one another. Thus, while some of the old battle lines continue to be the front of conflict between evolutionists and creationists, the line between secular science and revealed religion has materially shifted.

Notes

1.   Stephen J. Gould, *Rocks of Ages* (New York: Ballantine, 1999), 102.
2.   Ibid., 105, 106.
3.   Marx W. Wartofsky *Conceptual Foundations of Scientific Thought* (New York, Macmillan, 1968), 298.

4

The Reductive Temptation

While modern science is not necessarily reductive, it is the case that in its method, science begins by rejecting any principle of explanation which is not testable and whose effects are not measurable in a material way. Then it becomes easy or automatic for scientists, especially the more reflective ones, to assume that the scientific method is not only the best method for discovering truth but also that the universe is constructed in such a way that only science can rationally describe it. Thus, from method to metaphysics is the way science tends to proceed, i.e., from working principles about how to discover empirical causes for phenomena to the belief that the universe consists only of a system of material causes and effects. Individual scientists, however, tend not to generalize about scientific explanation, but instead they develop their reductive accounts as defined by the processes and entities fundamental to their own specialized field of scientific research.

Despite the tendency of scientists to write their philosophies out of the intellectual materials of their own specialized fields, a general sense of what reductionism entails regarding the relations among the different fields of science will be useful. The various fields of science can be arranged in a pyramid with the most basic science at the bottom and the least at the top. Following this model, the solid and secure base of the pyramid of scientific explanation are the so-called "hard sciences" of physics and chemistry including subfields such as particle physics and organic chemistry. Then resting on this great plinth are the middle level biological sciences such as cell biology and evolution, and resting upon biology is the apex of the pyramid, the social sciences including psychology, sociology, economics, anthropology, and possibly including humanistic fields such as history and ethics. Each of these three levels of the reductive pyramid, as well as the subfields enclosed within them, share some essence that defines them as scientific. Probably the best guess as to what that essence is, to the degree that all the fields and subfields claim to be scientific in the modern sense, is that they attempt to describe phenomena exclusively by means of material entities and physical causes.

Among the great scientists, the materiality of the universe is not a conclusion reached unreflectively but explicitly understood as the basis of their research. This does not mean that such scientists as Galileo, Newton, Einstein, and Bohr are simply philosophical materialists since, in fact, none of them were. It does mean, however, that as they approached their research, they defined the universe in such a way as to provide intellectual coherence and support to their efforts. Galileo revived ancient atomism as basis for his mathematical physics to distinguish between those aspects of physical phenomena that were subjective, such as color and sound, and those that were objective, i.e., those that could be studied by measurement and mathematics such as weight and size. This distinction between "primary" and "secondary" qualities is especially useful when defending the Copernican system in which a distinction must be made between sensible appearances of the Sun traveling overhead and the scientific reality of the Earth spinning daily on its axis. For Newton, the universe was material but mathematical as well. It could be described by a system of differential equations, and what could not be thus described was passed over—for, as he said in regard to gravity, he did not offer an hypothesis to explain it in essence but would leave it that he described its effects by his famous law of universal gravitation. Einstein shared Newton's belief about what scientific explanations should look like (i.e., a set of differential equations) and developed a theory of gravity which explained the history of the universe expressed mathematically but in force fields rather than lines of mutual force. For Bohr and Heisenberg, who dealt with subatomic phenomena, the physical universe was a mixture of perception and probabilistic events whose relations had to be calibrated to reveal mathematical relations of hitherto unknown complexity.

This tendency to define the universe in terms amenable to their mode of research means that scientists tend to understand the workings of nature as a complex network of physical causation which, as the more philosophically inclined scientists claim, can be organized into a logical system by which eventually all the phenomena in the universe will be explained. Usually the central element in the causal net is that entity which the scientists himself is most interested in, has provided the most fertile ground for his research, and has garnered attention both within science and with the general public. One of the foremost contemporary examples comes from physics in the current search for a "theory of everything" or TOE as exemplified most recently in "string theory" with its multiple dimensions. In contemporary theoretical physics, the challenge is to construct a unified theory that combines the two major twentieth-century theories of relativity and quantum mechanics, but the problem is one of mathematical consistency, which so far has eluded the physicists. The empirical element in the mix is sub-atomic particles such as electrons and quarks, protons, mesons, and the rest of what has been called "the particle zoo." Physicist Steven Weinberg states categorically, "the reason that we give the impression that we think that elementary particle physics is more fundamental than other branches of physics is because it is."[1] Weinberg thinks that particle physics is more fundamental than all the other branches of physics and, indeed, all other branches of science, We can infer that for Weinberg, elementary particle physics is fundamental because the universe is made up, at its most basic level, of elementary particles.

A second contemporary example of a scientific field whose researchers promote it as a basic science is sociobiology. Sociobiology, also called "evolutionary psychology," combines evolutionary and genetic science with social sciences such as ethology and anthropology, and it utilizes mathematical modeling to explain the behavior of animals including human beings. In this view, human beings are definitively animal—for all aspects of human nature which distinguish them from animals are rigorously explained as results of evolutionary drives and genetic necessity. This includes mother love, political idealism, acquisitiveness, aggressiveness, romance, and religion. E.O. Wilson states in his magisterial work, *Sociobiology*, "One of the functions of sociobiology, then, is to reformulate the foundations of the social sciences in a way that draws these subjects into the Modern Synthesis [of evolution with genetics]."[2] Wilson includes in the social sciences the fields of psychology, sociology, and ethnography, but also political science, economics, and ethics. For sociobiologists, the reductive element is the gene, the biological equivalent of the atom, which carries exact characteristics from one generation to the next according to evolutionary influences.

Particle physics and sociobiology provide the two foremost examples of contemporary reductionism and consideration of them immediately raises an issue. A problem with reductionism arises upon consideration of the large number of reductive theories that have been offered over time by scientists and scientific philosophers from Hobbes to LaPlace, Comte, Marx, Freud, Monod, Dennett, and finally to Weinberg—for like the varieties of doctrines manifested by all the Christian denominations, they cannot *all* be right. In the contemporary situation, if Wilson is correct that all human behavior is reducible to social biology, than Weinberg's assertion of a reduction to atomic particles becomes problematic if it is intended to refer to human behavior. It might be claimed that sociobiology can be reduced to particle physics, but no one has ever attempted this Herculean task since the explanatory pathways have to descend from evolution by natural selection—a principle that in itself seems to be irreducible to cell biology, biochemistry, molecular biology, inorganic chemistry, solid state physics, or to particle physics. In any case, evolutionary biologists such as Wilson and Dawkins do not ever mention the reduction of their field to physics and write as if they are convinced that their field is the base science.

This conflict then raises the issue of which, among the sciences, is the base science, i.e., the science to which all others are reduced: biology or physics. Physics can claim that it studies the most basic constituents of reality, but biology can claim that it acts as the bridge between the hard sciences of physics and chemistry and the social sciences including psychology and sociology. The question of which field constitutes the base science is sharply debated among scientists themselves. (A famous biologist once stated portentously, "One day biology and physics will meet, and it will not be biology which loses.") Weinberg, a Nobel Prize winning physicist, debated this issue with biologist Ernst Mayr, an influential evolutionary theorist who claimed that certain biological principles of explanation were not reducible to physics. Even within physics, however, Weinberg entered a sharp debate with a solid state physicist who thought that the principles of particle physics were "in no sense more fundamental" than other scientific fields and that other scientific fields are not reducible to particle physics—including his own field which studies matter in its solid state![3] But again, the foremost contemporary rivalry for claiming that theirs is the base science is between physics and biology, and both Wilson and Weinberg have provided book-length expositions of their respective reductive schemas—reductionisms that are reflections of their particular fields.[4]

Attaining the ultimate reductive theory would bring science to the fulfillment of the greatest intellectual project in the history of mankind, yet the project seems to bring out a mingling of fearful dread along with hubristic awe and intellectual triumph. Weinberg has written that "the more the universe seems comprehensible, the more it seems pointless." As if the anticipated reductive triumph somehow is destined to disappoint and, indeed, as if existence itself is not enough to satisfy the hubris and vaulting ambitions of the scientists. But ultimate comprehensibility is also the province of religious belief which is why physicists such as Einstein and Hawking who search after the final theory often talk about the "mind of God" as if the scientific secret they are about to unveil is the godhead; they understood this as the answer to the ultimate question of existence. Replacing the revealed God with reductive science, however, may resemble the sense of sex inexperienced adolescents get from dirty books—a misleading and unfulfilling substitute for the real thing. The ultimate reductive theory would not be the discovery of final truth, but a final reduction of the mystery of being to a mere physical law, and so the final accomplishment of knowledge becomes an act of existential despoliation.

A more practical objection to the reductive project is that the many interconnections between the sciences that would tie together their various theories have never been fully described, indeed, in most instances not described at all. Within physics, the links between its two most important theories, relativity and quantum, remain undiscovered, while the links between chemistry and biology, biology with psychology or among sociology, psychology, and economics have only been tentatively described and remain major points of contention. The number of explanatory linkages required between two or more scientific fields may be so large as to be virtually infinite and so complex as to be beyond our ability to imagine, according to some philosophers of science. And always there will persist arguments made by the scientists whose fields are the targets of reductive takeovers that something is being left out, a local law or a holistic perspective that is unique to their field and inherently irreducible. In biology, the current takeover target for the physical sciences, there has been considerable success for the reductive agenda in areas such as genetics and DNA, and currently a cadre of behavioral scientists is providing sociobiological explanations for numerous kinds of animal and human behavior. However, none of the biological reductionists spends any time on demonstrating their fields' general dependence on the physical sciences—on how biology will be reduced to biochemistry, biophysics, organic chemistry, quantum chemistry, etc. Philosopher of science Alexander Rosenberg, writing of reduction from genetic to molecular laws, stated, "this sort of reductionism is not only methodologically useless, it is probably unobtainable by agents of our cognitive and computational powers. Reductionism thus seems fated to cast little light on intertheoretical relations in biology."[5] Loath as scientists are to admit it, it is likely that the intellects of humans cannot handle the truth.

*   *   *

There is a dark aspect of the otherwise noble pursuit to explain a complex set of phenomena to a basic scientific level in that its motive is not always the benign one of the search for ultimate intellectual harmony. There is an "in your face" quality to reductive analysis; as if the scientists want to say that all the common person loves or takes for granted as the basis for his life is false. The reductive scientists, it seems, cannot help but gloat as they demolish these expectations as the following examples demonstrate: Skinner reduced "freedom and dignity" to conditioned behaviors, Wilson reduced Christian charity to "altruism," Dawkins reduced all human behavior to the desires of needy genes, Laplace reduced free will to iron laws of physics, nineteenth-century evolutionists reduced politics to racial survival, and contemporary mind researchers such as Dennett reduce the unity of the human intellect to multiple uncoordinated "frames" within the brain.

The reductive temptation, as we are calling it, is an assertion of the power of science as it actually exists in society and as its reductive advocates would like it to be. There is a megalomaniac quality to it as seen in the blatant assertions of reductionist scientists and philosophers who manifest a Nietzschean "will to power" in their attempts to explain all phenomena (but human phenomena especially) to whatever particular field of research they happen to favor. In this, the reductive temptation leads to a destructive tendency that proclaims the project of Enlightenment science is complete only at the point that all sense of human importance is lost. Reductivists assert that since the universe is infinite, the Earth and the Sun merely average examples among trillions of astronomical bodies, the number of organic species so large, and the nature of man so obviously animal, that human beings do not count for very much in the universe, if for anything at all. The reductionists say that the protests of religious critics emanate from fear of the truth. Religious believers must hide behind their faith because they are unable to cope with the reality of meaninglessness that is at the center of the universe–the black hole at the heart of existence. This portrayal leaves the scientist in the unlikely role of existentialist hero, as if only he has the courage to face reductive reality, and, like Rick and Captain Renault in the final scene of "Casablanca," the scientists walk into the fog of uncertainty having proudly done their duty in confronting the agents of religious oppression and bigotry.

But is the danger from reductive takeover schemes by ambitious scientists really all that much of danger? It may be asked, and further it may be said that while it is true that an uncompromisingly reductive stance applied to religion has destructive effects, it is also the case that much contemporary science and many scientists do not stand in adamant opposition to revealed religion. Indeed, the intellectual atmosphere is filled with various irenic schemes from advocates of both the scientific and religious fields which aim at healing the breach between the two. Typically, however, these irenic attempts assume the dominance of science, a tendency which can be seen in the manner in which philosophers have dealt with science. Philosophy since Kant and Bacon has deferred to science to the extent of looking to it for the fundamental model of knowledge, as if there was no other legitimate source of true knowledge. Many major philosophers, indeed whole schools of liberal, positivistic, or naturalistic philosophy have rejected the idea that humane learning, intuition, dialectic, and revelation provided sources of knowledge independently of the methods, theories, and discoveries of modern science.

Circumstances have changed since Kant's time, however, for modern science no longer wins universal applause. Instead, science these days is often looked upon with suspicion because despite the advantages of a technologically advanced life style, after three centuries of discovery and application science now presents us with as many problems as solutions. New sources of energy bring pollution of the air and water, experimental knowledge of the atom brings atomic weapons, super-efficient manufacturing techniques bring about unemployment, advanced medical technologies become so expensive they are unaffordable, electronic communications make it easy to invade privacy, and so forth. Scientific advancements, therefore, are now evaluated on a cost-benefit basis and are no longer seen as an unalloyed "good thing" or as a harbinger of certain improvement in our lives.

Nor is science in the twenty-first century perceived to move ceaselessly forward in the theoretical sense, especially in physics, which provides our understanding of the basic features of the physical world. The fascinating developments in physics that have occurred in the last hundred years, including the discovery of radiation, electromagnetic waves, the Big Bang, relativity, quantum indeterminacy, heat death, black holes, and quarks, have also constituted a radical reformulation of how science understands the physical universe. It has not been just one revolution in physical theory, but a series of theoretical revolutions which has made it far less plausible to assume that any scientific theory is final. Today, to say that no scientific theory is final is a cliché, but in the two centuries between Kant in the eighteenth century and Einstein in the twentieth, it was believed that modern empirical science, and only science, could achieve finality in its explanations and statements. That the epistemological dominance of modern science is no longer so widely accepted is a major renovation in our cultural thought patterns—a fact that gives heft to the religious and humanistic rejection of the reductive temptation.

Notes

1. Steven Weinberg, *Dreams of a Final Theory* (New York: Pantheon, 1992), 58.
2. Edward O. Wilson, *Sociobiology, The Abridged Edition* (Cambridge, MA: Harvard University Press), 4.
3. Weinberg, *Dreams*, 58.
4. Weinberg, Ibid. 58. Edward O. Wilson, *Consilience* (New York: Vintage, 1998).
5. Alexander Rosenberg, *Instrumental Biology or the Disunity of Science* (Chicago: University of Chicago Press, 1994), 22.

5

Kuhn's New History of Science

The Massachusetts Institute of Technology supports, along with its scientific research programs, a graduate program in science writing. The brochure that explains the features of this program states the following:

> Science writers address the larger public about the science and technology that shapes modern life, as well as the broader social issues—nuclear power safety, for example, or bioethics, or the environment—that science so profoundly influences.
>
> Science writers respect scientists and engineers, but don't treat their work as privileged, or as immune from informed criticism.
>
> Science writers never forget that science exists within a human and historical frame—and supply their readers with context as needed.
>
> Science writers may, or may not, hold academic credentials in science or engineering. But they are always humanists.

The existence of such science writing programs, a fairly new phenomenon, is an indicator of how important the social and other ramifications of scientific progress are in society and reflects the tremendous interest of the general public. That the science educators at MIT, arguably the institution most representative of scientific and technological progress, are aware of these facts concerning the public response to science indicates their importance. But what is more important are the words and concepts used to address the question of what "science writers" do, for as recounted in the brochure, their job is not just to describe science according to the way that scientists and engineers themselves do but to criticize it in a humanistic and historical context. The postmodernist intent of the program is thereby signified in that humanists—not scientists and engineers—will decide on the true value of scientific research and its applications. What may not be appreciated at first is how radical this view is and to what degree it represents a virtual revolution in the general public's understanding and in the willingness of intellectuals to accept scientists' statements about the significance of their work without scrutiny. Many different aspects of science and culture can no doubt be discerned as causes for this change in the attitude toward science, but one of the most significant is the intellectual influence of the philosopher and historian of science Thomas Kuhn, who finished his career as an MIT professor.

Thomas Kuhn has passed into the Western intellectual canon and his main ideas are related in rote fashion as much as those of any classical philosopher. "Revolution," "normal science," "anomaly," and "paradigm" have become familiar terms as is his notion that science is as much a sociological and historical as a theoretical or experimental enterprise. However, Kuhn's famous work, *The Structure of Scientific Revolutions,* is not only a case for the sociology of scientific knowledge or a book about paradigms, it is a book about the history of science, how that history develops in a certain manner with a certain rhythm, and, above all, how understanding that history compels readers to understand the nature of science in a new and different way. That the historical point of view was critical to Kuhn's analysis of scientific revolutions is indicated by the initial chapter entitled, "Introduction: A Role for History." Kuhn writes, "History, if viewed as more than a repository for anecdote or chronology, could produce a decisive transformation in the image of science by which we are now possessed."[1]

Where did Kuhn glean his essential insight about how the history of science could transform our understanding of science, we might ask. The answer is, at Harvard as an undergraduate, then as a graduate student, and then finally as a professor. Kuhn trained as a physicist and received his doctorate in that most scientific of scientific fields, physics. Usually seen as the "base science" upon which all other sciences rest, its methods and theories are taken as the iconic image of what true science is. James Bryant Conant, who had been head of the Chemistry Department, was president of the university, and it was Conant's influence that directed Kuhn toward the history of science. Conant had advised Truman on the use of the atomic bomb and the development of atomic power and so was well aware of the powerful influence that scientific and technological development had on society. As president of Harvard, Conant dealt with a severe pedagogical problem: *given that science and technology have assumed such an important, indeed, decisive role in our lives, how can we transmit what science is to undergraduates in the humanities who have no interest and perhaps no ability in science?* Conant's answer was to promote a required program of education in the history of science that used the case history method to present episodes of scientific discovery. The students read original reports of the experiments and from them lessons about the nature of science were explicitly drawn. (Subsequently, a book containing these cases was published entitled *Harvard Case Histories in Experimental Science*, 1950.)

Among the young professors that Conant recruited to teach the course was Kuhn who subsequently devoted his career to studying and writing about the history of science rather than pursuing physical research. However, his view of science history went beyond Conant's. Kuhn's view implied that the history of science was a means by which historical and broadly humanistic learning could not only understand science, albeit within the limits prescribed by an assumed lack of mathematical facility, but could compass science by subsuming it within the general rubrics of historical understanding. After Kuhn, the educated public for the first time might assume that science could be understood without actually knowing such things as the differential calculus, chemical equations, or the experimental method. Science's emplacement as a human activity made it knowable by its history, or as knowable as really mattered.

But what was distinctive about the new history of science which, as Kuhn said, was giving a new insight into the nature of science itself? On a technical level, Kuhn answered by stating that the more historians of science studied, for instance, "Aristotelian dynamics, phlogistic chemistry, or caloric thermodynamics, the more certain they feel that these once current views of nature were, as a whole, neither less scientific nor more the product of human idiosyncrasy than those current today."[2] As a result, the new historians of science no longer accepted what can be called the "Whig history of science," in which discoveries and ideas accumulated to result in an increasing body of certain scientific knowledge. This was essentially a triumphalist view of science history that tended to look at those episodes where new scientific ideas were opposed by the church, by inherited tradition, or by academic philosophy but which eventually triumphed as in Galileo's controversy with the Pope over the Copernican system. In the triumphalist view, certain scientific theories which had been overthrown, such as Aristotle's theory of motion in which bodies moved only when they were subject to force and the phlogiston theory which described heat as a chemical substance, were foolish mistakes caused by the scientist's prejudice or insufficient understanding of proper scientific method. However, Kuhn decided that these theories, although subsequently discarded, were no less properly scientific than the competitor theories that replaced them.

Kuhn cited a newly developing concept of science history which discarded the Whig history and attempted to give proper understanding to each episode of scientific discovery on its own terms—citing the work of Alexander Koyre in particular.[3] However, Koyre, a student of Heidegger's, was not the only such historian since over fifty years prior to Kuhn's research scientific histories had been appearing which implicitly or directly contradicted the view that science proceeded from victory to certain victory, from hypothesis to verified truth. Two scientists who worked in the period from the late nineteenth to the early twentieth centuries deserve mention. The prominent German physicist Ernst Mach wrote a history of mechanics, published in 1883, entitled *The Science of Mechanics: A Critical and Historical Account of its Development,* which was not a story of accretion of empirical truths physicists had discovered about motion but was meant to highlight the uncertain situation that physics faced in the late 1800's. Mach's history was an extended criticism of the foundations of Newtonian mechanics and an argument that mechanics was not the basis of all other physics and of science. In the same period, the French physicist Pierre Duhem, while doing original research in the archives of French universities, discovered that prior to Galileo and Leonardo daVinci medieval scholars had done extensive work on the development of a theory of motion. What his research showed was that Galileo had predecessors, such as Buridan and Oresme, and that the arrival of modern science was a matter of development not revolution. Interestingly, Koyre explicitly and Kuhn implicitly disagreed with Duhem's account and later attempted to re-institute the sharp distinction between the Christian, Medieval, and modern scientific worldviews. Koyre's book on the subject is entitled *From the Closed World to the Infinite Universe* (1957), while Kuhn's first book (written prior to *The Structure of Scientific Revolutions*) was *The Copernican Revolution*—both titles indicating vividly their disagreement with Duhem's "continuity thesis."[4] Thus, by the time Kuhn took up his own research, the history of science had become a field on which varying accounts and contrasting agendas contested.

Important works of science history by other scholars appeared in the first two-thirds of the twentieth century. This includes a series of pieces published in one volume as *The Origins of Modern Science 1300—1800* (1957) by the respected English historian Herbert Butterfield, and *The Idea of Nature* (1945) by another English historian who is also an underrated philosopher, R.G. Collingwood. Works by George Sarton, an American, should also be mentioned and, in particular, another book by an American scholar originally written as a doctoral thesis while at Columbia University, *The Metaphysical Foundations of Modern Science* (1952) by E.A. Burtt.[5] Burtt's volume gave detailed accounts of the philosophical, as opposed to the experimental, aspects of the thinking of the founders of modern science including Galileo, Kepler, Hooke, Descartes, and Newton. In effect, these early great scientists were philosophers as much as experimental scientists, a point which is taken for granted these days but which was surprising in the 1930's when Burtt first published. Now familiar with hermeneutics and critical theory, we do not find it exceptional that Einstein received a two volume treatment in the "The Library of Living Philosophers." But until the 1950's really it was held that there was such a sharp difference between science as the foremost means of knowing the universe and any other form of knowledge such that express efforts were made to overcome the gulf between science and the humanities, most notably in Snow's *Two Cultures*.[6]

Science history was presenting in these works, first the connections between science and other forms of human activity such as philosophy and politics and second, evidence that theories which had been dismissed were not necessarily less "scientific" (indeed they sometimes made comebacks, as did the wave theory of light). Finally, in effect, science could be looked at from the outside as much as from the inside, i.e., as much from the history of its ideas and their ramifications as from the actual discoveries and the mathematicized theories which were used to explain them.

Kuhn is sometimes seen as a philosopher of science in the tradition of logical positivism and the Vienna Circle who somehow went astray or whose account of paradigm change has a certain plausibility when applied to large changes of scientific worldview, such as the Copernican revolution, but which is implausible when applied to the way science is usually done. These criticisms of Kuhn are given from the point of view of twentieth-century philosophy of science imbued with a basic philosophy of naturalism, empiricism, and rationalism. The appeal of Kuhn's theory, however, cannot be explained as a sort of failed empiricism but rather as a theory that transcended empiricism because it gave an account of the nature of modern science beyond empirically defined methodology and scientific logic.

The place to go to measure Kuhn's philosophy of science is not the Vienna Circle or comparisons to the "hypothetico-deductive method" but to an older philosophic tradition known as the "philosophy of history." And here we leave the scientist's laboratory with its measuring devices, computer models, and brightly-lit workbench covered with precision instruments as the model of scientific research activity, and move to the scholar's library. Here the shelves hold centuries' old volumes of research reports from classical experiments, letters, and old journals, and books of ancient and modern philosophy wherein the larger patterns of the history of science can be discovered.[7] Philosophy of history in the grand, speculative manner as done by Vico, Hegel, Spengler, and Toynbee worked from the details of certain defined periods in history to discover patterns and repeated rhythms. These philosophic historians also tended to divide up history into certain ages and epochs, described as "worldviews," which defined the ideas, perceptions, and experiences of their time. Kuhn had done much the same thing; limiting his research to what interested him most and what he knew most about. He concentrated solely on the history of science and discovered a rhythm of scientific discovery in the alternation of periods of normal research under the aegis of a paradigm, which resulted in the eventual accumulation of "contradictions" with the arrival of a new paradigm to replace the old by means of a scientific "revolution." The paradigms for which Kuhn is famous he described as a kind of master theory of such power and comprehension that they literally defined a scientific field.

Despite the resemblances, however, Kuhn's view of the history of science was not intended as a "philosophy of science history" in the traditional sense, as can be seen in his general reaction to criticisms of his historical account of science and in what Kuhn himself left out of his theory. In response to critics, Kuhn relied on further clarifications which did not satisfy them, and in the end he pared down the range of what paradigms comprehended from large theories to small ones, thereby limiting their applicability to the history of science and the nature of science as a whole.[8] However, the larger and more comprehensive view of the applicability of paradigm theory is accepted by later commentators including physicist Steven Weinberg, even though he is a strong critic of Kuhn's historical account of science.[9]

Kuhn's response to his critics was in effect a retreat: for it appears that at base Kuhn's own philosophic commitments was an assumed naturalism that is often associated with scientific discovery and the scientific view of the universe. This was a philosophy that Kuhn largely shared with his critics within the community of philosophers of science at the time that left him no intellectual defensive line behind which he could take a stand against his critics. The best defense ultimately would have been a good offense—one based on the history of science as a constitutive essence of science itself. In order to do this, however, Kuhn would have had to develop further his own historical view of science as a full-blown philosophy of history, and here it can be seen that Kuhn's account is a philosophy of history of science by accident, not by intent. Kuhn left out the origins of science from his account, providing no overall view of the history of modern science from its origins in seventeenth-century Europe until the twentieth century. By that time the scientific/technological paradigm had come to dominate not just a large part of the thinking and life of the West, but literally of the entire world. Such an account would have had to deal more thoroughly with the development of the social sciences and not just the hard sciences in order to explore the interaction and possible cause and effect relationships between the scientific method and the objects of scientific study.

Some of Kuhn's critics perceived a political message in his account of science, particularly in his account of "normal science," i.e., that longer period of accumulated experiment between paradigmatic revolutions that generally characterized how science was done. Some of these critics, including Popper and Feyerabend, imputed to Kuhn an agenda of normalization of dogmatic attitudes, a defense of normalcy and convention, and a placid acceptance of the status quo. To the defenders of a heroic view of science such as Popper, Kuhn appeared to have elevated the normal periods of science above the times of paradigmatic change and to have degraded the heroic image of the scientist as an inventor of daring new theoretical constructs. Kuhn, they said, implied that the real story of the history of science was nonrevolutionary—consisting of a group of researchers working diligently in their laboratories or research settings, doing normal science which meant being intellectually satisfied while filling out the paradigm. As Kuhn described it, they were "puzzle solving"—a term that seemed to have particularly provoked the critics. Feyerabend hinted that there was a fascistic tenor to Kuhn's elevation of normal science, stating that he was unable to accept Kuhn's "general *ideology* [which] could only give comfort to the most narrowminded and the most conceited kind of specialism. It would tend to inhibit the advancement of knowledge ... [and] increase the anti-humanitarian tendencies which are such a disquieting feature of much of post-Newtonian science" [italics in original]. Feyerabend asked portentously, "Is it his [Kuhn's] intention to provide a historico-scientific justification for the ever growing need to identify with some group?"[10] Thus, for both Popper and Feyerabend, the more significant criticism came against what they saw as Kuhn's elevation of the group over the individual. However, another element not necessarily attached to this inherently political criticism entered into Popper's response, namely, his belief in the heroic nature of the true scientist who battles religious bigotry, the ignorance of the common people, the institutions of government and, the conventional thinking of his own scientific colleagues.

Kuhn's politically sensitized critics assumed that his description of what science is like historically is prescriptive as well as descriptive in intent—an assumption especially apparent in Stephen Fuller's recent book *Thomas Kuhn: A Philosophical History for Our Times*. For Fuller, Kuhn is the cause of the bad situation of contemporary social thought, which has turned away from its critical vocation to that of a support for the status quo—for "Kuhn's 'acritical' perspective has colonized the academy."[11] Kuhn, as Fuller explains it, as the protégé of James Bryant Conant, was caught up in the web of institutional complacency and inherent conservatism of the postwar American ruling class. In effect, Kuhn's influence is a part of American prosecution of the Cold War and American antipathy toward progressive reform on the political plane, and it supports suppression of critical modes of thinking on the academic plane. In this criticism, Fuller, who calls himself a "devout social constructivist," reflects the attitudes of the Frankfurt School including critics such as Theodore Adorno.[12] However, it is the politics of the critic that is uppermost rather than Kuhn's since Fuller, as one of his reviewers pointed out, belongs to that school of left-wing critics for whom no subject is ever "radical enough." His politics support a permanent intellectual revolution which refuses to rest in any intellectual position or to look benignly on any institution, rather it regards them all as not merely suspect but inherently coercive.

In using Kuhn's description of science history as a scapegoat for their fears, Kuhn's politically minded critics must be credited with an intuition of something real. The implied assertion that science and scientific method are unexceptional in comparison with other modes of human activity including art, literature, politics, and religion, discredits one of the major themes of the Enlightenment, namely, the theme that human reason is sufficient for a comprehensive understanding of the universe and that by its own efforts mankind can progress to a new and elevated form of existence. What Kuhn's view of science discredits politically is the Enlightenment belief in human progress and human perfectibility—a belief which underlies both socialist and free-market ideologies. It is notable that politicized criticism of Kuhn's account of science comes from left-wingers (Fuller), Libertarians (Popper), and anarchists (Feyerabend). Here is the real intersection between Kuhn's historical theory of science and a form of conservative politics. While it is improbable to suggest that Kuhn's theory intended to support the bland establishment politics of the Eisenhower administration or looked forward to the free-market policies of the Reagan administration or to the neo-conservative foreign policy of the second Bush administration, it does support a philosophical brand of conservatism. That is, Kuhn's historical philosophy of science denies the Enlightenment premise that human agency is capable of transcending normal human fallibilities and limitations—a lesson usually learned through a consideration of the travails of human history with its failures, tragedies, evils, misunderstandings, and repetitions but is now apparent from the history of science.

In his politicized understanding of Kuhn's philosophy of science, Stephen Fuller criticized Kuhn's theory because it had left something undone and had not fulfilled what Fuller claimed was its main purpose, namely, arriving at a "critical" view of science and instead supported a view of science conformable with establishment politics. In this way, Fuller was explaining his own view of the relationship that should occur between modern science and contemporary civilization; this was a view based on the assumption of a displacement between the ideals of science as it should be done (critically) and the fact of science as it is actually done (conventionally). That Fuller discerns a tension between science and the conventions of Western society is a contemporary insight more typical of the twenty-first than the nineteenth century, or the first half of the twentieth when science was seen as the vanguard of progress of Western civilization. Fuller denies this interconnection between progress and science, seeing science's proper role now as that of the base of critical evaluation of society's lack of progress according to left-wing standards. However, Fuller's criticism is a photographically negative image of the true failures of Kuhn's account of science. Kuhn did not go far enough in his emplacement of science in the historical context of Western civilization or of humanistic learning and failed to provide a coherent account of the relation between an historical view of science and truth.

The true conflict between science and conventional society is whether the underpinnings of our civilization should be shredded by scientific reductions of its high ideals be they political, philosophical, and religious, or whether modern science should be defined as a part of Western civilization. In effect, the conflict is whether the ultimate outcome should be that history (as representative of the humanistic traditions of Western civilization) should be reduced to a scientific account of material forces and a naturalistic understanding of human nature (e.g., sociobiology), or if it should be understood not in terms of the success of its latest theories but developmentally, in historical terms as a part of Western civilization. In short, the debate is if it is to be the science of history or the history of science. Kuhn raised the question and succeeded in convincing many intellectuals that the historical view of science is the proper one and is the most effective way of understanding modern science. The tremendous growth of historical and cultural works about science in the last twenty years, including whole schools of critical philosophy and a flood of semipopular books about science, is a fulfillment of the implied agenda in Kuhn's historical view of science.[13]

In this way, Kuhn is the most influential of the postmodern philosophers, of far more importance in permanent influence and resonance than Foucault, Lyotard, or Derrida. The historicized vision of science that is now current is deprived of its status as the primary truth teller of Western culture and frees up the imagination to consider religious as well as humanistic alternatives. As a result promoters of the religious ideal have much less work to do than in any time since before Darwin.

Notes

1.   Thomas Kuhn, *The Structure of Scientific Revolutions* (2nd ed.), (Chicago, University of Chicago Press, 970), 1.
2.   Ibid., 2.
3.   Ibid., 3.
4.   Pierre Duhem, *The Aim and Structure of Physical Theory*, translated by P. P. Wiener (Princeton, NJ: Princeton University Press, 1954). Reprint, (New York: Atheneum, 1977). Alexandre Koyre, *From the Closed World to the Infinite Universe* (Baltimore, MD: Johns Hopkins Press, 1957). Thomas Kuhn, *The Copernican Revolution* (Cambridge, MA: Harvard University Press, 1957).
5.   Herbert Butterfield, *The Origins of Modern Science 1300–1800* (New York: Free Press, 1957). R. G. Collingwood, *The Idea of Nature* (New York: Oxford University Press, 1945). Edwin A. Burtt, *The Metaphysical Foundations of Modern Science* (Garden City, NJ: Doubleday, 1954).
6.   Paul A. Schilpp, ed., *Albert Einstein: Philosopher–Scientist, Library of Living Philosophers Vol. VII* (Chicago: Open Court, 1949). C. P. Snow, *The Two Cultures* (Cambridge: Cambridge University Press, 1993).
7.   There are numerous representatives of the philosophy of history who can be cited, but for a concise statement see R. G. Collingwood, *The Idea of History*, edited by T. M. Knox (Oxford, Oxford University Press, 1956) 1–13.
8.   See the section "Normal Science," in *Criticism and the Growth of Knowledge*, edited byt I. Lakatos and A. Musgrave (Cambridge: Cambridge University Press, 1970), 249–259.
9.   Steven Weinberg, "The Revolution That Didn't Happen" in the *New York Review of Books, Vol. XLV, #15* (1998). See also a volume of essays about Kuhn entitled *World Changes*, edited by Paul Horwich (Cambridge, MA: MIT Press), 1993.
10.   Lakatos and Musgrave, 197–199.
11.   Steven Fuller, *Thomas Kuhn: A Philosophical History for Our Times* (Chicago: University of Chicago Press, 1970), xv.
12.   Ibid., xvi.
13.   Ibid., xv.

6

Galileo's Enduring Career

Galileo Galilei stands as the iconic figure in the war between science and religion. The battle lines have shifted from when the war first started, and nowhere is this more apparent than in the reputation of Galileo. Galileo's career has endured well beyond his own lifetime and our understanding of it has undergone various transmutations—for it contains such interest as it affects a number of different aspects of the relation of science with religion and of both to culture.

There are, it seems, many Galileos: Galileo as avatar of freedom of knowledge against religion and bigotry; Galileo as a clever and ambitious careerist who overtopped himself when he tried to trick the Pope; Galileo as atomist philosopher; Galileo as one of the great scientists of the "new physics" second in his accomplishments only to Newton; Galileo as victim of thought police who convinced him to admit his guilt by psychological pressure and by threats of physical torture, comparable to an old Bolshevik held prisoner in the Lubianka; Galileo a prisoner of conscience turned on by the authority whose thanks he merited like Oppenheimer and Sakharov; Galileo first among scientists to sell the products of his research to support further scientific research; Galileo the first but surely not the last physicist to adapt his new knowledge to the uses of war and to sell his research results to the military; Galileo disturber of social peace; and recently, Galileo as father to an intelligent and perceptive daughter in Dava Sobel's account which also gives an overall picture of Galileo's life and work.[1] Galileo's career endures because all these varying and conflicting interpretations have been believd and are suggested by elements plainly present in his career.

Galileo is a seventeenth-century figure and his career begins in the late Renaissance at a time when religious wars were taking place and Protestantism was established in England and northern Europe. What was developing in Protestant countries was a solemn brief against the Catholic Church. Protestantism defended itself against the charge of wrecking the unity of the Christian faith by denigrating the Catholic Church and elevating the right of the individual believer over the doctrines of the church.

In this context of religious warfare, Galileo's insistence on promoting what many Catholics and Protestants thought was a heretical doctrine was an irritant to both sides—an issue seemingly irrelevant to the major crisis of the day but one that brought into sharp relief the issue of faith and reason. Added to the religious context was the secular context of the growth of capitalism and of a newly enriched and empowered middle class that was challenging the old feudal political arrangement of lord and serf. This was a new arrangement which encouraged the belief in any individual's ability to succeed and find his place in life by the application of his own intelligence to the problems and challenges of his own time and circumstances. In this combined religious and secular context, the case of Galileo had a deep resonance because of his condemnation by the Catholic Church, his time as a prisoner of the dreaded Inquisition, and his ultimate vindication in the central matter of the Copernican theory—the episode which, in effect, marks the beginning of modern science.

In a view held for a long time by many people, religious bigotry had done its worst with Galileo, reducing him to a terrified old man abjuring his life's work done on behalf of the enlightenment of the human race. But this act of oppression of truth by ignorant faith is punished by the vindication of Galileo's central premise: the Earth does indeed travel around the Sun, and not vice versa—a fact known now to schoolchildren and itself is so much a matter of common sense that to deny it is a paradigmatic example of stupidity. Religion in the social form of the Catholic Church, which pronounced against science and reason and science in the figure of Galileo, had eventually been victorious. After his conviction and abjuration, Galileo became a hero while living in house arrest, using his remaining time (Galileo was an old man at this point) to conduct experiments and write serious scientific works in the form of dialogues. Galileo's house became a stop on the tour of the continent that young Englishmen took, John Milton and Thomas Hobbes among them. Visiting the old intellectual fighter whose career was a vindication of the new science and of freedom of thought was a part of their education after their formal studies were completed. Galileo as hero of science and victim of religious oppression constitutes what may be called the "classical" or standard view of the Galileo controversy that was expressed in a sumptuous presentation on public television and a coffee table book in 1974 entitled *The Ascent of Man* by Jacob Bronowski. Bronowski said, "there was never any doubt that Galileo would be silenced, because the division between him and those in authority was absolute. They believed that faith should dominate; and Galileo believed that truth should persuade."[2]

There were details, however, which when presented by serious historians or Catholic apologists, made it apparent that the story of Galileo was much more complex. For one thing, Galileo was, by the time of his ultimate conflict with the church, a celebrity scientist, well known throughout Italy and Europe for his astronomical researches using the telescope, his enunciation of newly discovered physical laws that he put in precise geometrical form, as well as his controversies with philosophers, preachers, and other scientists. Galileo had an aggressive, humorous, and vivid personality which made him welcome in the highest circles of society. Thus, when a new pope was elected, a man of learning and sophistication who also traveled in high circles, he and Galileo were well known to each other. Yet it was this pope, Urban VIII, who thought he had an understanding with Galileo and who when he discovered otherwise, he took revenge by putting Galileo's case before the Inquisition.

It was the nature of this "understanding" with Urban VIII, and Galileo's clever use of it, that was the immediate cause of the controversy. The major issue was the course of the planets and the Sun in relation to each other; these were the two competing "world systems,"[3] as Galileo called them, the Copernican versus the Ptolemaic accounts. It was this scientific controversy that had become a major social and religious issue, which stands as the classic case of a conflict between secular and revealed knowledge. The Bible had statements that made it clear that in the view of the bible writers, the Sun made a daily circuit about the Earth and not the other way around. However, the texts in which this assumption occurred were not connected in a vital way to revealed doctrine. The church's attitude toward the Bible was that even though it was a revealed document, every text of the Bible was not to be taken in the most literal sense; certain texts of necessity had to be interpreted metaphorically or spiritually. However, the earth-centered interpretation of the biblical texts was reinforced by the philosophy of Aristotle, who was influential in the universities of the day. His philosophy depended upon the astronomy of the ancient astronomer Ptolemy, which was earth-centered. In Aristotle's vision, the Earth was the central sphere in a system of nested spheres identified with heavenly bodies including the Sun, the Moon, the planets, and the stars.

Besides contradicting certain biblical texts and Aristotle's philosophy, the Copernican doctrine contradicted the universally accepted "naive" notion of earth centeredness experienced by every competent human being on the planet earth. This was the idea that, in fact, the Sun does travel about Earth, from sunup where it appears at dawn in the east, to noon where it is approximately overhead, to sunset where it sets in the west. The inevitability of this fact arises from the immemorial experience of the human race and resistance to the Copernican view was no doubt based on the perceived circular motion of the Sun. To accept the Copernican system a person had to exercise a feat of imagination. He must picture that he lived on a huge ball, which was turning on its axis while at the same time hurtling though outer space around the Sun in a combined double circular motion that somehow neither made objects on the Earth fly away into space, nor made the Earth itself crack up and disintegrate. Galileo had answered these and other such objections in his *Dialog on the Two Systems*, the book that got him into such trouble with Pope Urban VIII and the Inquisition. The reason for the trouble was that in their earlier discussion about Galileo's projected book, the pope had argued that the truth regarding the two world systems was ultimately beyond final answer; since physical issues could never be settled, ultimate truth was available only through religious revelation accepted on faith. Galileo seemingly agreed, and the Pope therefore expected that the dialog would set off the two world systems in opposition to each other without a declared winner. Instead, the total effect of the dialog was to vindicate the Copernican system and logically dismantle the Ptolemaic, thus violating what the Pope thought was the explicit agreement he had with Galileo. (Einstein writes that Galileo was "down-right roguish" in this regard.[4]) However legitimate the Pope's anger, all the more increased because only a close reading of Galileo's long and dense work would reveal its Copernican agenda, his response was plainly an overreaction. Thus, despite his trickery, Galileo became the victim if not the hero of the story.

*   *   *

We live in a time when hero worship is uncommon, where heroes are looked upon with suspicion, and where revisionist accounts are written to show up heroes as not being very heroic. The same thing happened to Galileo as has happened to Thomas Jefferson or Harry Truman. As an example, Bertold Brecht's play, *Galileo*, is in effect an accusation of cowardice based on Galileo's not standing up to the Inquisition rather than abjuring the Copernican doctrine in order to save his liberty and his life. (Brecht, who lived under the protection of the East German Communist regime, would not seem to have good standing to make such an accusation.) On a scholarly level, Galileo's reputation as an avatar of the freedom of scientific thought has been subject to revision from a variety of sources in the last fifty years, starting with the publication of *The Crime of Galileo,* by Georgio de Santillana, in 1955. Historians have become more aware of the complexities of the situation that Galileo and the church were in at the time and more willing to understand the point of view of the church—not as a matter of apologetics but as a matter of trying to better understand the events as they took place. DeSantillana, not a Catholic, could go so far as to evince sympathy for the Roman Inquisition that tried Galileo. Comparing his trial to the witch trials taking place in Boston and to twentieth-century Russian Communist show trials, he writes, "We must, if anything, admire the cautiousness and legal scruples of the Roman authorities in that civilized period."[5] DeSantillana's point is not to vindicate the church's trial of Galileo, but to understand it apart from the stereotype of dark ecclesiastical villains versus the single-minded defender of scientific truth. A less condemnatory attitude toward the church's role in Galileo's case can also be seen in an account by historian of science Charles C. Gillespie. He argues in 1960 that the conflict arose not because of irresolvable doctrinal differences (the church already having in place a procedure for dealing with conflicts between biblical texts and proven scientific facts) but for personal and political reasons. "The drama between science and the Church, therefore, unfolds with that inevitability which is tragic because it arises from the characters of men rather than the necessity in things."[6]

A further impetus to Galilean revisionism comes about from scientific, rather than historiographic, developments—that is, from considerations of Einstein's theory of relativity. Relativity means, among other things, that there is no preferred observer and no place in the universe such as a center from which absolute measurements of physical phenomena can be made. In one presentation of relativity, Einstein compared his theory of relativity to what he imputed to "Galilean relativity."[7] Here Einstein was referring to the manner in which Galileo compared relative motions and how they might be measured, for instance, while looking at a ship at sea. On the ship, a sailor walks back and forth on the deck. He can calculate the speed of his own pace, while an observer on shore sees the motion of the sailor as a compound of his motion relative to the ship plus the motion of the ship relative to the shore. For Einstein, there is no final place from which to determine the motions of the sailor or the ship since the shore sits on the Earth, which moves relative to the Sun, which moves relative to the galaxy, etc. Such considerations are fatal for the Copernican debate which enmeshed Galileo and the church if the issue is the absolute motion of the Earth and the Sun, since absolute motion does not exist in the relativistic universe. Philosopher Karl Popper summarizes the relativistic view with characteristic accuracy, even though it is a view he disagrees with.

> In support of the view that Galileo suffered for the sake of a pseudo-problem it has been asserted that in light of a logically more advanced system of physics Galileo's problem has in fact dissolved into nothing. Einstein's general principle, one often hears, makes it quite clear that it is meaningless to speak of absolute motion, even in the case of rotation; for we can freely choose whatever system we wish to be (relatively) at rest. Thus Galileo's problem vanishes.[8]

Its worth noting that among those who thought that Einstein's relativity theory made the issue between Galileo and the church moot, was the late Astronomer Royal of England, Fred Hoyle.[9] The relativistic issue goes beyond the range of physical theory to the question of who was right in terms of scientific methodology—the pope, who asserted to Galileo that the Copernican issue could never be resolved, or Galileo, who thought that the result could be made definite. Here it seems that Urban VIII wins the argument, for he took the position not that the earth-centered system of Aristotle and Ptolemy was physically true because the Bible and the philosophy professors said so, but rather that the issue could never really be resolved, like other learned religious believers of the time who considered the Copernican issue including John Milton who expressed his view in his Christian epic *Paradise Lost*.[10] The issue of the relative motion of the Earth and the Sun did not appear, to the pope, to affect the central religious concerns of the Christian faith which is based on the salvific life of Jesus Christ. To deny Christ's resurrection would be to attack a central belief, but to assert that the earth traveled about the Sun (rather than the other way around) was tangential. In short, Christian doctrine could accommodate the possible or even likely truth of the Copernican system if it could be proved.

It might seem as if however much detail might be added to the historical record, that no serious commentator, especially a well-regarded philosopher of science, would ever think to reverse the general understanding of the famous incident of Galileo's conflict with the church. Nonetheless, Paul Feyerabend, speaking of Galileo's trial, says that Galileo was "treated rather mildly" but that "a small claque of intellectuals aided by scandal-hungry writers succeeded in blowing it up into enormous dimensions so that what was basically an altercation between an expert and an institution defending a wider view of things now almost looks like a battle between heaven and hell."[11]

Feyerabend was not a religious believer or a defender of the Catholic Church or religious belief generally. Instead, his judgment reflects a view about scientific truth that was radical in the context of the Anglo-American philosophy of science and the positivism of the Vienna Circle, but which is increasingly accepted today. Namely, that scientific truth has no more credibility or validity than truths reached by any other means. Feyerabend explains that scientific truth has always been subject to controls and influences from outside institutions. This can be seen today when drug companies keep their research results secret because they may be lucrative or when the department of defense regulates access to research done under its auspices (for example, in refusing access to encryption software deemed necessary for the national defense). In the Galileo case, the scientist had been instructed by church authorities that he could hold the Copernican theory only as a hypothesis until he had actual scientific proof—proof which was lacking at the time. Also, as we have seen, Urban VIII's opinion about the inability to decide between the two competing doctrines is now more credible in light of Einstein's theory of relativity. Thus, Feyerabend concludes his discussion of Galileo: "To sum up: the judgment of the Church experts was scientifically correct and had the right social intention, viz. to protect people from the machinations of specialists. It wanted to protect people from being corrupted by a narrow ideology that might work in restricted domains but was incapable of sustaining a harmonious life."[12] Feyerabend was a left-wing anarchist in his politics, a position which harmonized with his belief that no one social institution, including science, had a market on truth. In Feyerabend's treatment, Galileo becomes an example of the kind of expert whose doctrines are deleterious to the body politic and who should be constrained, rather than a hero of thought.

Ironically, thirty years after Feyerabend presented his argument that the Catholic Church's actions against Galileo had been well justified, Pope John-Paul II weighed in on the case in 1992. He did not condemn Galileo but apologized and corrected the record by admitting to all those interested that Galileo's conviction in the Inquisitorial trial had been wrong. This exoneration came four hundred years after the case but is still welcome. It was apparently part of John-Paul II's general agenda as leader of the Catholic Church to apologize to various groups offended over the previous centuries including, most appropriately, the Jewish people. Pope John-Paul II's statements were generous and irenically motivated, and it must be supposed that the relativistic interpretation of the case would let the church off the hook was deemed not relevant or was not understood. In fact, even accepting that Urban VIII's idea of scientific explanation was superior to Galileo's, this does not exonerate Urban VIII or the church from guilt for the manner in which Galileo was treated. Some scholars think that evidence was faked for the purposes of convicting Galileo, i.e., that unknown church officials manufactured a letter written sixteen years earlier by Cardinal Bellarmine which instructed Galileo not to hold the Copernican theory *in any way,* not even as a hypothesis.[13]

All the interpretations of his career which have multiplied in recent times can be seen as a reflection of the fact that Galileo is the proto-scientist. He was the first modern scientist in his methodology and his ruthless elimination of non-material causality in the physical universe. He was the first to come in full and fatal conflict with religious authority (not excluding Giordano Bruno whose doctrines were less scientific than magical and pantheistic), the first to hint at a world system expressed fully in mathematical terms, and the first to sell his research. The scientist came to exemplify both the positive and negative qualities we now associate with modern science. Thus, it is only fair to give Galileo the last word, one that is particularly useful for a discussion of the contemporary conflicts, real and otherwise, between religion and science. At one point prior to his arrest and trial Galileo sent an open letter to a Princess Christina, a dowager of the powerful Medici family, this was not a personal letter but a long argumentative tract of over fourteen thousand words. Galileo had been accused by a preacher in her court of attacking the Bible by asserting doctrines that contradicted sacred scripture. Galileo naturally wanted to respond in order to protect his reputation but also because such an accusation could get him involved with church authorities investigating heresy (which eventually, of course, did happen). By this time Galileo had been able to support the Copernican theory with many experimental and mathematical pieces of evidence. Now his enemies, rebutted by his arguments, were reduced to attacking the Copernican theory as "damnable and heretical" because it appeared to contradict various biblical passages—including the account where Joshua makes the Sun stand still until the Israelite army completes a mop-up operation (*Joshua, 10: 12–13*)

Galileo defended himself by putting forward an extended argument that the Bible should not be used as a text to decide scientific issues since it often happened that the biblical writers had to explain things in terms that could be understood by ancient Israelites, Greeks, Romans,

Abyssinians, etc. Galileo's position regarding the use of biblical texts in scientific research is widely accepted among Christian theologians today, but was itself based on the traditional teaching of the church authorities including Bonaventure, Aquinas, and Augustine. In effect, scientific knowledge was sometimes required to interpret biblical texts in order to decide whether texts were to be interpreted metaphorically or "morally" rather than literally.[14]

> Therefore, I think that in disputes about natural phenomena one must begin not with the authority of scriptural passages but with sensory experience and necessary demonstrations. For the Holy Scripture and nature derive equally from the Godhead, the former as the dictation of the Holy Spirit and the latter as the obedient executrix of God's order … God reveals Himself to us no less excellently in the effects of nature than in the sacred words of Scripture … and so it seems that a natural phenomena which is placed before our eyes by sensory experience or proved by necessary demonstrations should not be called into question, let alone condemned, on account of scriptural passages whose words appear to have a different meaning.[15]

Galileo's warning about the misuse of sacred scripture is particularly apt at this time when sincere but wrongheaded believers make statements about the age of the earth based on certain biblical texts. The Bible was written to reveal to human beings that which they otherwise might never know about creation, salvation, and the nature of the divine. The Bible was not written with the purpose of instructing us about the purely physical aspects of the universe—facts that as the Creator has arranged it, the human intellect is perfectly able to figure it out on its own.

Notes

1. Dava Sobel, *Galileo's Daughter* (New York, Walker, 1999).
2. Jacob Bronowski, *The Ascent of Man* (Boston, Little, Brown; 1973), 205.
3. Galileo Galilei, *Dialogue Concerning the Two Chief World Systems*. Translated by Stillman Drake (University of California Press, 1967).
4. Ibid., xi.
5. Georgio DeSantillana, *The Crime of Galileo* (Chicago, University of Chicago Press, 1959), 228.
6. Charles Gillispie, *The Edge of Objectivity* (Princeton, Princeton University Press, 1960), 48.
7. Albert Einstein and Infeld, L., *The Evolution of Physics* (New York, Simon and Schuster, 1961), 158.
8. Karl Popper, *Conjectures and Refutations* (New York, Harper and Row, 1965), 110.
9. Fred Hoyle, *Nicolaus Copernicus* (New York, Harper and Row, 1973).
10. John Milton, *Paradise Lost* (many editions); See Book VIII, lines 66–140, Angel Rafael to Adam.
11. Paul Feyerabend, *Against Method* (New York: Verso, 1988), 13. See also Wade Roland, *Galileo's Mistake: A New Look at the Epic Confrontation* (New York: Arcade, 2003).

12.   Ibid., 137.
13.   DeSantillana, *Galileo,* 125–131.
14.   Saint Augustine. *On Christian Teaching,* translated by RPH Green. (New York, Oxford University Press, 1999), 4.
15.   Galileo Galilei, "Letter to the Princess Christina" in *The Galileo Affair: A Documentary History*, edited and translated by Maurice A, Finocchiaro (Berkeley, CA: University of California Press, 1989), 92.

7

Why the Physicists Speak of God

The point of these essays on the science and religion "wars" is not to make a case about which side should win, but to paint a picture of what is going on in the war. In the present instance, the point is not to say that physicists *ought* to be talking about God or that the study of physical nature leads inevitably to the reality of God, but simply to describe the current situation. What is going on is that physicists are talking about God, a lot, and have been doing so at least since the 1980's, and they are talking about God in a more specific way than many theologians and philosophers. One British physicist (a Christian) was recently quoted as expressing sympathy for humanists in the matter of God: "Our dear friends in the humanities do get themselves awfully confused about whether the world exists, about whether each other exists, about whether words mean anything. Until they have sorted out whether cats and dogs exist or not, or are only figments in the mind of the reader, let alone the writer, then they are going to have problems talking about God."[1] It is a statement that carries a nice air of assurance and an implied criticism of postmodern critical thought. However, while it is true that physicists are talking and writing about God, they are not agreed on what to say about whether He exists or not, and if He does exist, in what fashion, i.e., as a person independently of our intellects or as an impression of the rationality of the physical universe. The reasons why the physicists speak of God have less to do with a religious revival among research scientists than the mode and content of the scientific theories discovered during the twentieth century—in what can be called postmodern physics.

That the physicists are talking and writing about God is a fact that can be easily demonstrated by a quick mention of recent books. For example, Stephen Hawking's bestseller *A Brief History of Time* for which the late celebrity scientist Carl Sagan wrote in the introduction. "This is also a book about God ... or perhaps about the absence of God. The word God fills these pages."[2] *The Science of God* by an Orthodox Jewish physicist attempts to demonstrate the "convergence of scientific and biblical wisdom." *The Fire in the Equations* described as "a fascinating discussion of scientific discoveries and their impact on our beliefs" including the topic of "how God might answer prayers" from the point of view of physics, is not written by a physicist but reviews recent discoveries including the "unreasonable effectiveness" of mathematics. *The Way the World Is* is the first in a series of books on the topic by John Polkinghorne, who is both an accomplished physicist and an Anglican clergyman, which attempts a synthesis of traditional Christian doctrine and the vision of the universe rendered by postmodern physics. Two seminal books in this vein had appeared earlier, one of which was *God and the New Physics* (1982) by physicist Paul Davies. This book gave a detailed but readable account of the new discoveries, such as the Big Bang and quantum mechanics, and opened up the possibility of religious belief while keeping a neutral stance on the reality of God.[3] *God and the Astronomers* (1978), by NASA scientist Robert Jastrow, made the case that the Big Bang was the same event described in *The Book of Genesis* when the Lord created the universe. Jastrow wrote of the discovery of the Big Bang in a quotable statement: "The scientist has scaled the mountains of ignorance; he is about to conquer the highest peak [but] as he pulls himself over the final rock, he is greeted by a band of theologians who have been sitting there for centuries."[4]

Other books connecting God and physics can be cited, but two other well-known books which do not explicitly mention God but which deal with other aspects of postmodern physics are worth mentioning as they indicate that physics, since the early twentieth century, has become impressionistically speaking less positivistic and materialistically. Two physicists, John D. Barrow and Frank J. Tipler who studied under the physical cosmologist John Wheeler, produced a massive tome *The Anthropic Cosmological Principle* that describes in detailed historical and scientific terms the "anthropic" principle, various forms of which are used by some physicists to explain the existence of life and consciousness in the universe.[5] The book attempts to reintroduce a strong teleological element into contemporary science based on the evident improbability of life and consciousness ever having evolved otherwise. Another well-known volume, which has been reprinted several times, is *The Tao of Physics* by Fritjof Capra. This book argues "that modern physics leads us to a view of the world which is very similar to the views held by mystics of all ages and traditions." Capra writes further that, "Modern physics has confirmed most dramatically one of the basic ideas of Eastern mysticism; that all the concepts we use to describe nature are limited, that they are not features of reality … but creations of the mind."[6]

Physicists talking about God has become a virtual publishing phenomenon, but the reason why this has occurred is what gives meaning to the phenomenon. Physics is the "base science"—that field of modern science with the most prestige because it deals with the most general aspects of physical existence. Time, energy, matter, and force are the physical substrata of all that we see in the inorganic universe including earth, air, fire, water, minerals, rocks on the earth's surface, and the earth itself. The entities described by physics underlie the myriad transformations of matter, all chemical processes, and the operations of all living things. The connections between physics and chemical and biological phenomena are made via subfields such as biochemistry, biophysics, chemical physics, and physical chemistry. Physics has always had this appeal—that in understanding it one could understand the most basic facts of physical existence and virtually of all reality. As such it compelled the interest of ancient figures including Socrates and Augustine, both of whom, however, left the study of physics represented by such thinkers as Thales and Anaximander for the study of human beings; this was a subject they found even more compelling and more complex. This is in contrast to the contemporary physicist Steven Weinberg who reported that he was at first a philosophy major as an undergraduate but switched to physics in his search for the ground of ultimate reality.

Popular interest in modern physics was apparent at the time of Newton when refined ladies discussed calculus in eighteenth-century salons and has recently been rekindled in the book reading public as a result of the profound discoveries that took place in physics mostly during the first half of the twentieth century. It is these discoveries, associated with the names of Einstein, Bohr, Shrodinger, Heisenberg, and Planck, which shook up physics to the degree that the assertion that physics supports the kind of materialism that was influential in the nineteenth century is no longer plausible. It is also these discoveries which led some physicists to talk about God in the context of their own field. It is an important phenomenon because it is a reversion to the past when such early giants as Galileo, Kepler and Newton found support for their Christian beliefs in the discoveries that they found in the night skies. The stars, for them, announced the glory of the creator God rather than, as later scientists have asserted, a vast volume of nothingness which announced that man and his desires were of no ultimate significance.

The reason why the physicists speak of God is revealed by the series of physical discoveries that permanently and profoundly redefined how science views the most general aspects of the physical universe. Four recent physical discoveries will be briefly described here from the point of view of their import for religious belief: the Big Bang, quantum indeterminacy, the nature of matter, and anthropic principles.

The Big Bang

We often become disappointed in our childhood heroes when, as we grow up, we learn of their all too human deficiencies. In my middle school years, I learned of the humility of the great Einstein; he had been informed by a mathematician that he had made a simple algebraic error regarding a minus sign in one of his relativistic equations, and he had written to thank the man and then corrected the equation. As a grown man I learned that while the story was true, a Russian mathematician named Friedmann had in fact notified Einstein about a minus sign, the matter was not trivial but literally of cosmic importance, and Einstein had only reluctantly conceded Friedmann's point. The issue was complex for it involved the shape of the universe; in his set of relativistic equations, Einstein had used a constant of zero (0) and of a positive one (+1), each constant giving a differently shaped universe (in four dimensions). What Einstein had not considered was the constant of a minus one (-1) and the kind of universe using a constant of minus one throughout his set of equations would describe. As it happened, the minus one constant produced a model of a dynamic universe that expanded from a point in time and space and exploded outward. This is the model of the universe that physics has settled on as most accurately reflecting the extensive body of scientific evidence that we now have about the shape and origins of the universe.

But why had Einstein declined the suggestion, refusing at first even to acknowledge Friedmann's letter? Why did Einstein refuse to acknowledge that his relativistic theories produced a model of the universe that would later be vindicated by scientific evidence and that described the "Big Bang" which is probably the most important scientific discovery of the twentieth century? There were two reasons, the first was that Einstein was somewhat conservative in his physics and was not sympathetic to the idea of an expanding universe, preferring a static, unchanging universe, such as the one Newton had described. The other reason, according to Robert Jastrow, was that this version of the universe implied that it had a beginning in time, and that the first mention of such an event was in the Bible.[7] What was true of Einstein is true of many physicists since then; they are very uncomfortable with the Big Bang because it implies that the origin of the universe is beyond what physics can know (i.e., before the Big Bang when time did not exist), but also because it gives aid and comfort to the religious enemy. According to Stephen Hawking, "Many people do not like the idea that time has a beginning, probably because it smacks of divine intervention. (The Catholic Church, on the other hand, seized on the big bang model and in 1951 officially pronounced it to be in accordance with the Bible.)"[8] Physicists have subsequently developed many alternative hypotheses to explain the origin of the universe, but the Big Bang prevails at least as a first order approximation based on all the available evidence. Thus, whether they approve or disapprove of the idea that the universe had a beginning in time or that the Big Bang implies a Creator, the physicists are still talking about God.

Quantum Indeterminacy

There is literally "no room for God" in a deterministic universe, which is why atheism and science were seen as conjoined during the nineteenth century—the heyday of deterministic physics. When Laplace famously told Napoleon that he had no need for God as a hypothesis, Laplace was also saying that in the new mechanical universe described by deterministic equations there was no *room* for God. Divine agency simply did not fit in since it was understood to be an irruption into the mathematical laws and physical processes by which the universe ran. God was not only not needed, but divine agency was a direct challenge to scientific understanding. The nineteenth century was the century of scientific determinism, due to the coherence and confidence of the account that the physicists presented which tended to leak into the general culture. Great systems of deterministic explanation, including those of Marx, Spencer, and Freud, were erected and have had an unfortunate and enduring influence. In large part this was because they replaced religion with a presumably more rational and up-to-date scientific account of the universe. In these systems, it was not so much that scientific knowledge overcame revealed religion as that science assumed the role of religion since Marx's scientific socialism, evolutionary naturalism, and Freudian psychology provide ethical, psychological, and spiritual guidance to millions to this very day.

The plausibility of determinism in physics and of the deterministic quasi-scientific systems of the nineteenth century rested on a picture of the physical universe as a machine—as an automatically running system that needed no external guidance or divine agency, yet would produce the marvelous effects of human consciousness and the progressive civilization of the Victorian era. Human knowledge was assumed to be limited strictly to secular scientific knowledge and the conclusions of the scientists assumed to be completely objective. All this was about to change, however, largely due to the fatal discovery of quantum indeterminacy.

Quantum indeterminacy is fatal to the deterministic worldview because it proclaims that at the heart of the atom, scientific knowledge is suddenly and irredeemably limited. Atoms themselves were held to the ultimate element of physical matter by ancient philosophers as well as modern scientists, and it was conceptually disconcerting for scientists to discover at the start of the twentieth century that the atom was made up of even smaller particles—electrons, protons, and neutrons (the list of subatomic particles was due to expand exponentially). But the problem didn't end there for then developed a contradiction that physics lives with to this day. It was discovered that deterministic laws could not predict the motions of atoms and subatomic particles, this means that the Newtonian calculus was useless at this most basic level of physical reality and that the laws of probability had to be substituted instead. This implied that there was an inherent limit to what scientists could ever know about the smallest and most basic elements of the physical universe, the subatomic particles that make up the atom. It also implied that what scientists really knew was not the objective reality of atomic particles, but their relation to the subatomic particles. Relying on the mathematics of probability is like betting on a horse race for the physicist. Like a horse race, before an atomic collision the results are merely probable, and only after the race can the results be known with precision and the winners paid off. But then the nature of the description of the atomic collision or the race depends on when the observer knows about it; if before, the outcome is probable at best and may be simply unknowable; if after, then the observer's or the bettor's knowledge is definite. It all depends on when the observer learned about the event. It appears that what the physicist can know about subatomic reality is not the determined course of particles, but only ranges of probability.

The final and inescapable implication of quantum indeterminism is that the physical universe is not a machine, which renders much less plausible the notions that all physical events are predetermined, that men have no free will, and that God does not relate in any meaningful way to the physical universe. As a result, we have seen a large decline in belief in the great deterministic systems of the nineteenth century and a renewed interest in the operations of the human mind, and, most interesting of all, a renewed interest in the relation of the creator God to the physical universe among physicists themselves as well in the general culture.

The Nature of Matter

Matter is the most important thing that physics studies since it is the most fundamental tangible aspect of physical existence. Matter has always been something of a mystery, and the ancient philosophers, such as Thales and Anaximander, explained it in various ways. Aristotle based his philosophy of nature on a categorical distinction between matter and its form, that is, on the physical stuff of existence versus not just its physical shape but also its intellectual reality, such as its purpose or function. Aristotle also thought that matter was eternal, that it had always existed and always would. Other ancient philosophers, such as Democritus and Lucretius, subscribed to an atomic theory declaring that the universe consisted only of atoms in motion in the void. They gave a modern sounding account of how things decay or grow, change shape and color, or are hard or soft by explaining such phenomena as the difference in the type of atoms and their relative positions.

The atomic theory is in a way a source of comfort to scientists and materialists, because it provides a background noise for the materialist intellect—a steady hum of assurance that the world can ultimately be explained as no more than atoms and the void. The presumed eternality of atoms gives reassurance as well—a sense of ultimate permanence no matter what other exigencies take place or what strange beliefs the common people indulge. This comfortable assumption was part of nineteenth-century thinking but was rudely exploded in the twentieth. Atomic matter was previously assumed to have existed forever entirely independently of human thought and the atoms themselves thought to be impenetrable. None of this is true, however, as in postmodern physics matter can wink in or out of existence. Atoms are not the most fundamental element but are made up of particles, the particles in turn are convertible into energy and into other particles. The number of subatomic particles is so large and their convertibility so complex that manuals are necessary for the experimenters to keep track of them. All this means that matter is no longer a clear competitor to the realm of the spirit since matter is evanescent and not eternal; it exists in a contingent and impermanent state as the particles flit in and out of existence. But if the physical world is no longer solid and no longer providing assurance of its eternity and stability, then the intellect turns to what underlies matter—another level of existence or the realm of spirit. God as creator is no longer a competitor to material existence but a complement and, perhaps, the source.

Anthropic Principles

The physical universe supports life, in particular, human life. This is self-evident from the fact that humans are the ones making this observation and since scientists are human, the fact of the existence of human life is sometimes interpreted as a *scientific* fact. Scientific facts differ from ordinary facts because they have to be explained in a certain way, and thus the reality of human life has to be explained by integration into a paradigm of scientific understanding. Understood scientifically, what becomes apparent is the inherent improbability of human life and life itself in its biological sense. The idea that the arrival of life in the universe is inevitable was plausible when the physical universe was understood to be a machine for the production of life—a notion reflected in popular science fiction which is filled with stories about strange and exotic forms of intelligent life on other planets and galaxies. But the arrival of life including its most primitive bacterial forms is now understood to be most implausible. In fact, life is nonexistent on other planets, such as Mars, where we have (so far) been able to look. When looked at objectively, the various stages through which the universe and the earth processed in order to bring about human life are extraordinary in their complexity and, scientifically speaking, improbable in the extreme. (See chapter 10)

An additional kind of explanatory principle seems to be required to explain the existence of intelligent life, the so-called "anthropic" principles which state that the universe is constructed in such a way as to bring about human life. The sense of the improbability of human life is reinforced by the existence of such physical properties as the electrolytic properties of common water (thought to be essential for any form of life to exist) and the Carbon atom's ability to form complex molecular chains, which is at the basis of all biochemistry. Then there are such arcane facts as the ratio of the weight of electrons to those of protons and the exact temperature which matter reached during the first three minutes after the Big Bang. These are inherent physical properties of matter which had to have existed precisely as they do to the millionth of a measure to support the existence of human life.[9] Whether a separate anthropic physical principle is required to explain why and how these facts and relationships exist in such exquisite harmony so as to support life is hotly contested among physicists, but the facts themselves are not, i.e., the physical properties and ratios. To discuss these issues is to discuss the why and how of human existence and inevitably to bring in discussions of cosmic purpose or the lack of it, and the existence of God or His nonexistence—which in any case leads the physicists to speak of God.

In the turn from nineteenth- to twentieth-century physics, there came about what may be termed a revolution in the way that physicists viewed the universe. The change from a universe infinitely the same for all time and in all directions, a monochromatic and vast void whose silence terrified Pascal, to an outer space filled with gravity waves, black holes, and bright galaxies, a universe which has a beginning in time and an unknown future, has had the effect of releasing the human intellect from a kind of prison. From condescending declarations by scientific philosophers who told the ignorant populace that atoms and mathematical forces explained it all, to a new sense of possibilities which include alternate universes, anthropic principles, indeterminacy, and a dynamic not a static universe, leads to the possibility that God exists in a signified relation to the universe and not merely as an outside observer. Which is why, whether they want to or not, the physicists are talking about God these days.

Notes

1. Tim Radford, "'Science cannot provide all the answers:' Why do so many scientists believe in God?" *The Guardian*, (September 4, 2004), [on-line].
2. Hawking, Steven, *A Brief History of Time* (New York: Bantam, 1988), x.
3. Gerald L. Schroeder, *The Science of God* (New York: Broadway Books, 1997). See also Kitty Ferguson, *The Fire in the Equations* (Grand Rapids, MI: Eerdmans, 1995). See also John Polkinghorne, *The Way the World Is* (Grand Rapids, MI: Eerdmans, 1983). See also Paul Davies, *God and the New Physics* (New York: Simon and Schuster, 1984).
4. Jastrow, Robert, *God and the Astronomers* (New York: Warner, 1978), 105–106.
5. John Barrow and Frank Tipler, *The Cosmic Anthropic Principle* (New York: Oxford University Press, 1998).
6. Fritjof Capra, *The Tao of Physics* third edition (Boston: Shambhala, 1991), 19, 160.
7. Jastrow, *Astronomers*, 17.
8. Hawking, *Time*, 46, 47.
9. Barrow and Tipler, *Anthropic*, 560–570.

8

The Dissolving Gene

Reference to the gene has become a primary feature of recent evolutionary rhetoric, but even in less rhetorical and more scientific presentations, the gene has assumed a central role. Thus, the very first section of the first chapter of E. O. Wilson's paradigm text, *Sociobiology,* is entitled, "The Morality of the Gene." This work explains not only Wilson's grandiose plan of integrating all the social sciences (and the humanities) into evolutionary biology, but also states the centrality of the gene in evolutionary theory. Reference to the gene, it seems, is required in order to explain "how can altruism which by definition reduces personal fitness, possibly evolve by natural selection?"[1] His answer, and that of contemporary social biology, is that natural selection: the engine of evolutionary change acts on genes not on the individual organism. The original Darwinian notion was that the origin of species was caused by struggles among separate individual organisms fighting for their own survival. However, this idea was contradicted by the observation that organisms sometimes acted in a way that put them in danger of death in order to save their offspring. Moreover, how could self-preservation explain the existence of insect casts such as drones that could not reproduce; this was a problem that particularly bothered Darwin. However, if organisms can be understood to act not to preserve themselves but their genes instead, then the self-sacrificial actions of organisms to protect their offspring and the existence of nonreproducing casts of insect can readily be explained. Thus, reference to the gene allows contemporary sociobiology to solve the issue of self-sacrificial actions and to expand evolutionary theory to account for the behavior as well as the physical form of living beings.

It can be argued that evolution as the primary theory of biology and as the fullest scientific explanation of living beings is being displaced by genetic science. The irony here is that genetic science which has been seen by biologists as the theory necessary to complete Darwinian evolution is now becoming the "senior partner," so that evolution is defined as a genetic process, rather than the other way around. E. O. Wilson defined natural selection as "the process whereby certain genes gain representation in the following generations superior to that of other genes located at the same chromosome positions."[2] In popular culture, the power of the word "gene" is such that it has become a totem word used by political scientists, social commentators, and psychologists to explain virtually every aspect of human existence from personal characteristics to family life to religion. This is due in part to the human genome project, the new technologies including DNA identification in criminal cases, gene alteration as a medical technique, and amniocentesis to provide evidence of heritable fetal disorders. But also, on a theoretical level, genes are now used to explain intelligence and behavior, and such explanations are now becoming common outside the precincts of biology. Because its deterministic aspect denigrates human freedom and the influence of social norms, the use of genetic explanations of social behaviors has become an inflammatory political issue.

The writings of Richard Dawkins provide the most prominent example of genetic rhetoric since he is unafraid to take the logical implications of genetic evolutionism to their conclusion, stating that all significant behaviors, especially human behaviors, are determined by genes. He goes so far as to say that the primary focus of biology in the study of life is not on individual organisms but on the genes, implying that it is softheaded to retain an atavistic belief in the biological reality of organism including perhaps human beings.[3] Dawkins further draws the atheistic theological implications of evolution to their fullest. While previously he had merely announced his atheism, lately his rhetoric has gone full bore against religion in general. His criticism follows a new pattern of attacks on religion, such as those by Sam Harris, in which religion is identified with its most extreme and cruel manifestations by using the terror tactics of Islamic fundamentalists as the final reductive element of what constitutes the essence of religious practice. In Dawkins' view, religion is not only inevitably a part of the human genome, it leads inevitably to oppression and terror. Such a position puts Dawkins in an untenable place, since it is unlikely that weakening the religious belief of Christians and Jews, which is Dawkins' immediate purpose, will in any way prevent the attacks of Muslim fanatics on Western society or weaken the belief of Muslims.[4] It is evident from his writings, as from those of his compatriot in the Darwin wars, Daniel Dennett, that evolution constitutes a philosophy and a guide to life; it is a replacement for revealed religion, politics, family ties, or any other ideal of life and society that might contend with the evolutionary world view. In his stream of evolutionary presentations, what Dawkins is most notable for is the reliance on genes as the ultimate element in evolutionary process and as the central entity in the understanding of life both in its abstract and existential modes. Dawkins' presentations reflect a widely held view among evolutionary biologists that genes are the ultimate reductive entity.

How did the "gene" achieve this status of totemic word by which proponents conjure explanations for everything attributable to life forms including shape, behavior, and intelligence? The "gene" accumulated its power because it works as a prime example of scientific causality; it is empirically discernable, theoretically indispensable, and mathematically clear. Despite its usefulness as a mode of scientific explanation, however, we are at liberty to analyze the gene in terms of its historical development, by means of philosophical analysis, and by its rhetorical and wider attachments. We are not compelled, after all, to lay down our intellectual arms, despite the blandishments of Wilson or the demands of Dawkins that we acquiesce to their account of the totemic powerful gene. Surprisingly perhaps, a sustained intellectual analysis of the concept of the gene will result in its dissolution from the ultimate reductive physical entity to a mixture of concepts and flow in the stream of biological process and scientific discovery.

Physically, genes are entities observable under an electron microscope that are known to directly control certain aspects of every organism's shape, size, and even behavior by means of complex but increasingly understood chains of biochemical reactions. That is, genes manufacture enzymes ("one gene, one enzyme"), which in turn manufacture proteins that are the building blocks of different kinds of living tissues; these tissues perform exact functions (including behavioral) within living organisms. It is their presumed physicality, their reality as observable physical things that persist over time, which gives the sense of their importance. These are the *real* causes of life forms, their shape, the functions of their internal parts, their behavior—or so it is assumed.

Ontologically, genes are often incorrectly understood to be whole and entire unto themselves like Leibniz'z *monads* or the atoms of chemistry. Originally, as scientists first dealt with them and as the public largely continues to accept them, they were thought to have no internal parts. Genes, however, exist as a kind of mediate entity halfway between cellular tissue clearly observable as the essential constituent of any living being and the molecules and atoms that we cannot observe, yet we take on the authority of science as real entities. The comparison of genes with atoms, however, points out a peril for the clarity and coherence of the concept. Atoms, which at first were taken to be without internal parts and an ultimate explanation of physical activity up until the late nineteenth century, were subsequently discovered to have internal parts which terribly complicated physicist's understanding of elemental physical reality. They went from the clear Newtonian vision of slightly elastic atomic particles following determinate laws of mathematics, to the "zoo" of particles and forces within the atom that could only be described by the mathematics of probability. The concept of the gene has undergone the same decomposition, for beneath the reality of the gene lie the constituent parts of the DNA chains which support the multitudinous combinations of the four chemical bases—C, G, A, and T.

Mathematically, the singularity of each gene means that genes take their place in formulas, which can be then used to predict their behavior and their effect on the larger aspects of the organisms that they are assumed to control. This might seem to mean that a kind of deterministic calculus could be applied to each gene that could describe its flow of action via enzymes, proteins, and tissues, etc. However, like the molecules which make up a quantity of a gas (millions of which are in rapid motion and cannot be individually tracked), genes cannot be treated usefully by applying deterministic mathematics to each one individually but must be treated as a group. This means that while the action of single gene is unpredictable to a degree, since in the process of genetic transmission it might be mutated or otherwise deformed, that genetic behavior can be treated mathematically as a whole—as in the case of "genetic drift" within a population. The mathematics that Mendel applied to his pea plant experiments were, in fact, probabilistic since only some proportion of a particular characteristic was transmitted, for example, approximately one-fourth in the case of recessive characteristics.

Reductively, genes sit close to bottom of the middle tier of the reductive pyramid, which includes all of biology, on which rests the top tier of the social sciences and below which rests the bottom tier of the physical sciences (see chapter 7). Genes sit as the ultimate biological reductive entity and as ultimate entities beyond or below which no appeal is possible, like a verdict of the Supreme Court or the pope. As such, genes are often described in contemporary evolutionary biology as the entity on which natural selection works, that is, it is genes and not individual organisms or general populations that compete for survival or dominance, and it is here that incoherence of the gene concept becomes apparent. The fact is that genes themselves can be subject to reductive analysis in which they are understood as chemical complexes, that is, large chemical structures made up of complex but mathematically describable arrangements of big molecules.

Philosophically, genes constitute both a clear and distinct idea. Indeed, the clarity of the concept is what seems to matter most since so much appears to be explained by the gene, and it has become for some biologists, most notably Dawkins, a basis for a virtual philosophy of nature and of man. However, there is such a thing as a concept being *too* clear, as can be seen from the nature of the discovery of the gene itself. The origin of the concept of the gene begins with the monk Mendel, a high school teacher and administrator, working his famous experiments with pea plants in the monastery garden. The story is well known. Mendel, after extremely careful observation and experiment, derived the laws of inheritance including laws that described dominant and recessive characteristics. He deduced from his empirical evidence the existence of a fundamental unit of inheritance which determined the separate characteristics of each pea plant (and of all living organisms) which was later (in 1909) termed the "gene." The characteristics he studied and the plant on which he studied were well chosen by Mendel—for characteristics such as the height or color of a particular plant or whether the pea itself was rounded or withered are especially noticeable in the pea plant. This enabled Mendel to make clear observations, which in turn led a clear inference that an agent was responsible for each characteristic.

Mendel's experiments with the pea plants are almost too clear, for as historians and scientists have observed, no one has ever been able to reproduce them in full. The characteristics he studied are not as easily observable as he presented them in his two reports, and further genetic research has shown that some of the characteristics he studied are determined not by one single gene but by several. This fact is often obscured by the clarity of the concept of the gene, for a one-to-one correspondence seems natural at first blush—one gene for eye color, for instance. Yet skin color, which appears to be a kind of blending depending on the inherited genetic characteristics of the parents, is controlled not by one but by several genes. Otherwise, skin color would not over several generations of "miscegenation" become coffee colored rather than either white or black depending on which color was dominant or recessive. Thus, it has been suggested that somehow Mendel "cheated" on his experiments and that he knew in advance how they would come out.[5]

But this suggestion seems self-contradictory, for what text could Mendel have cheated from since there were no texts describing the laws of inheritance or the gene until Mendel wrote his own?

There are two alternative explanations for Mendel's success. The first is that he was a monk and as part of his religious training would have had ingrained in him the habit of care and deliberation in everything he did, nothing done carelessly or heedlessly, each task to be presented as a sacrifice for the Lord. Thus, he may have simply conducted his experiments in a far more careful and controlled manner than is usual. That he carried out his experiments is not really in question for his research reports are extensive. While his actual experiments have not been reproduced in detail, his general conclusions that are mathematically precise have been reproduced with many organisms, many times, and they are the unquestioned basis for genetic research and understanding of the transmission of somatic characteristics.

The second possibility is that Mendel's genius lay in understanding the living process of genetic inheritance not as a biological but as a physical phenomenon and that his thought patterns were Galilean, so to speak, looking for the greatest simplicity and mathematical clarity which would yield up that form of explanation which seemed natural and true in itself. To do this it is necessary to have "a feeling for the organism," so as to see into its reductive essence, i.e., the connection between the inner structure and the outer characteristic.[6] In this way, it may be true that Mendel saw the results of his experiments ahead of time and subconsciously forced or simply saw the empirical evidence as he knew it should be; genes were in control of somatic characteristics in those precise ratios of one-fourth recessive to three-fourths dominant. If this is the case, then it is again the very clarity of the concept of the gene that beguiled Mendel and indeed led him to a powerful truth about inheritance. It is a kind of clarity that, if it does not mislead, at least convinces the intellect to ignore the complexities and complications that are inherent in living things and the process of life itself.

The further history of the gene presents evidence of its strange and compelling place in the history of biology, as it is well known that at the time that Mendel published his discoveries (including the concept of the gene and the enunciation of the four laws of genetics) no contemporary scientists understood or appreciated them. Indeed, he was dead and more than thirty years had passed before three researchers, in 1900, independently discovered his publications and integrated Mendel's discoveries with their own research. How characteristics were transmitted from one generation to another and how they changed was the primary question that biological science was then dealing with, and Mendel's discoveries provided the breakthrough. But why had it taken thirty-five years for the significance of Mendel's work to be discovered? To answer this question, we can revert to the ontology of the gene as it is reflected in the history of microscopic accomplishments in genetic and cellular research.

One of the things eventually discovered after the microscope was invented by Van Leeuwenhoek in the seventeenth century was that living tissue is organized into discrete *cells*, but all that could be discovered at a primary level of magnification in 1838 was that cells existed as an omnipresent aspect of living tissue. Further development of microscope technology was necessary before it could be discovered that cells were not the same through and through, but had distinguishable parts within them, i.e., the nucleus. Subsequent development of the technique of staining cells revealed little intertwined lines within the nucleus, which were distinguished under the microscope by their color—hence "chromosome" from the Greek for "color." Only then was the true significance of the nucleus understood as revealed in the extensive and groundbreaking experiments of T. G. Morgan in the early part of the twentieth century. Finally, the chromosome themselves were discovered to be chains of genes, a discovery made by chemical and biological analysis but confirmed by electron microscope observation—a twentieth-century technological development. The point of this brief account is that Mendel made his discovery, prediction really, before microscope technology had advanced to the point where it could observe either chromosomes or genes. Mendel had leaped from the level of the cell, down past the level of the chromosome to the gene without the use of microscopes, staining techniques, or an extended knowledge of the data of genetic transmission. The discovery of the gene was a process of inference on Mendel's part, not of direct observation. His pea plants had told him all, but the price of his success is that, like Van Gogh or an avant-garde poet, his work was not recognized in his own lifetime. It would take biological science more than thirty years to catch up.

The reason for dealing with the chronology of the discovery of the gene is to point out a moral, namely, that genes present a confused portrait and are not the simple physical entities that current rhetoric makes them out to be. That is, genes have an ideal and abstract aspect as well as a physical and empirical one, and it is in trying to make the two accounts coordinate that the gene loses its rhetorical effectiveness and some of its scientific explanatory power as well. Thus, we can detect a variability of the concept of the gene even within Dawkins' own writings on the subject.[7] At first, genes are assumed to bear a direct physical relationship to highly defined bodily characteristics as in Mendel's detection of seven observable characteristics of the pea plant, such as color, height, and whether the pea is wrinkled or smooth. Mendel's apparent assumption was that each of the seven was determined by the action of a single gene, but the one-to-one connection does not in most cases apply. The examples of one gene defining a single observable characteristic in fact are pathological, that is, the misfiring of one gene causes such tragic genetic diseases as Down's syndrome, Hunter's syndrome, and Huntington's chorea.

The discoveries about genes that are coming in great profusion these days have incited a great interest within the general public—an interest excited to further levels by the promises made by biological scientists that the new discoveries will yield cures for many diseases, including Alzheimer's and spinal cord injuries. This excitement, in turn, yields the practical benefit of research grants for the scientists to proceed on an array of new fronts in genetic and biological research. However, as we learn more about genetic processes, we also learn increasingly about their complexity so that nowadays genes are dissolved into their DNA and then into chemical bases, but chemical reduction is not the end of the process of the dissolution of the concept of the gene. The end comes, as Dawkins explains, when genes are described as units of information, and the analysis is no longer by means of description of its physical and chemical aspects, but as a carrier of data from the parent to the child organism. The step from describing the gene as a physical entity to a unit of information has the consequence of making it an abstract entity, so defined because it can now be included under the rubrics of information theory. (Remember that information resides independently of a particular physical carrier just as the same message can be carried by the sound waves of an aural conversation, electrical impulses over a landline telephone, radio or satellite as electromagnetic waves, and as marks on paper as "hard copy.")

In one of his later books, *River Out of Eden*, Dawkins utilizes a metaphor just as he had in his books describing the "selfish" gene and the "blind watchmaker," this one taken from a phrase in *Genesis* to explain the "Darwinian view of life" to a popular audience. For Dawkins, the river of life is made up not of spiritual living water but of the biological essence of life. These are not genes, perhaps surprisingly. They are the next lower ontological level that constitutes genes, namely, the DNA spiral molecules. "The river of my title is the river of DNA," he states, so already we have left the idea of the atomistic gene, self-sufficient and integral, for genes themselves are made up of DNA molecules. But in the next step we leave the realm of the physical altogether; Dawkins explains, the river is a "river of information, not a river of bones and tissues: a river of abstract instructions for building bodies, not a river of solid bodies themselves. The information passes through the bodies."[8]

The informational aspect of the gene is reflected in the fact that Dawkins has invented a new term, "meme," which not only rhymes with "gene" but also attempts to give the same clarity to ideas that genes gives to evolution. Dawkins himself seems to have been surprised at the success of the new term, and references to "memes" now occur in semipopular writing that has nothing to do with evolution. The ultimate reason for the power of the gene concept is not scientific so much as rhetorical, or perhaps philosophic, for it is the gene as a reductive entity that is at the basis of its appeal. Exact features of the physical bodies of organisms and discernable features of their behaviors are said to be determined by specific genes or sets of genes, so that all the essential features of organisms are thereby explained. The process of reductively explaining phenomena of generational resemblance by means of hidden scientific rules has an enormous appeal to our intellects; it enables us to perceive the "real" explanation of things hidden under the skin of the common world. Just as the apparent motion of the Sun circling around the Earth is explained by our existing on the surface while it rotates on its axis, so also the complexity of organisms in the constant yet ever changing stream of life is explained by a mechanical principle (natural selection) working on integral, whole, singular particles—namely, genes.

The problem with ultimate reductive entities is that they are subsequently discovered to be composed of parts that then replace them as the "real" reductive entity; but by then the clarity given by reductive explanations is beyond repair. Atoms, once discovered as real entities by chemistry in the late nineteenth century when empirically analyzed in laboratories in the early twentieth century, were discovered as having parts such as electrons and nuclei, which obeyed their own set of probabilistic scientific laws not anticipated by scientists of the time. Once atoms were no longer seen to be solid and the same through and through, our understanding of physical reality became inordinately complexified, and, once split, the atom could never be put back together again.

What is currently happening to the concept of the gene parallels closely what happened in the early twentieth century to the concept of the atom. It has dissolved into its constituent parts, and further analysis reveals not some ultimate reductive entity—protons for atoms, DNA for genes—but rather engages the scientific mind into consideration of the limits of empirical knowledge and the means of acquiring exact data. Revealingly, both subatomic reality and DNA activity are expressed in the nondeterministic mathematics of probability. The reductive promise of ultimate physical entities has not been fulfilled and the rhetoric of genes is not supported by scientific fact.

Notes

1. Edward O. Wilson, *Sociobiology*, 3.
2. Ibid., 3. Also see chapter 4, "The relevant principles of Population Biology" and chapter 5, "Group Selection and Altruism."
3. Richard Dawkins, *The Selfish Gene* (Oxford: Oxford University Press, 1989), 237, ff.
4. Richard Dawkins, *The God Delusion* (Boston, Houghton Mifflin, 2006), 20–27.
5. R. A. Fisher, "Has Mendel's Work Been Rediscovered?" *Annals of Science* 1 (1936), 115–137. See also Jan Sapp, "The Nine Lives of Gregor Mendel" in *Experimental Inquiries* edited by H. E. LeGrand (Norwell, MA: Kluwer Academic Publishers, 1991), 137–166.
6. Gillispie, *Edge of Objectivity*, 333–337. See also Evelyn Fox Keller, *A Feeling for the Organism* (New York: W. H. Freeman, 1993).
7. Ullica Sagerstrale, *Defenders of the Truth* (New York: Oxford University Press, 2000), 137–138.
8. Richard Dawkins, *River Out of Eden* (New York: Basic Books, 1995), 4.

9

How Evolution Came from Outer Space

It is an implied but powerful aspect of evolutionary explanation that it is deterministic and that as a scientific law, evolution can explain every single significant aspect of human reality without miraculous intervention or teleological principles. The strongest indication of the implicitly deterministic aspect is that Darwin and all orthodox Darwinists until the present day agree that evolutionary change as species develop one from another by "slow and insensible degrees," such that the differences between successor species is minimal. In this way, evolution has the aspect of a gradual and machinelike process; its results are predictable and require, once again, no special additives to enhance the automatic process and produce its remarkable results. Also, the inherently probabilistic nature of evolutionary process is kept within a range that makes it reasonably predictable once the initial set of condition is spelled out.

The implicit appeal of the theory of evolution is that with one or two basic laws with attendant corollaries, the history of life on the planet earth can be fully explained. That is, in principle, evolution extends from the very beginning before life appeared on earth, to the rise of multiple forms of living things, to the current situation of mankind spread over the planet. In this, evolutionary theory fulfills the basic deterministic ideal of scientific explanation; given a set of initial conditions, the application of a scientific theory will predict in detail all the successive stages that follow the initial conditions.

In the case of the release of a gas into a volume containing gases of a different temperature, the rate of expansion and the eventual degree at which the temperature of the volume settles can be predicted by the application of well-known, empirically proven, and mathematically precise scientific laws. It is important to note that even though the microprocesses of individual molecules bumping into one another is unpredictable, on a macro-level and within a narrow range of probability, the process of gaseous expansion is predictable. This is so also in evolution; given an initial set of conditions, namely, the earth as it was approximately 4.5 billion years ago before life came upon it, evolutionary theory predicts the course of life from its very beginnings to the evolution of separate species that became increasingly complex. It moves from the sea where life began to land and from one-celled to multi-celled organisms—from plants, to fish, to amphibians, to land creatures, to mammals, and finally to man. Beginning with the first men who lived in tribal and primitive societies and moving to increasingly complex levels of civilization reaching, ultimately, industrial society. Ideally, evolution by means of natural selection will explain all this in the form of a deterministic set of biological laws, and even as the separate microprocesses may be unpredictable, the development of a particular species for example, the macroprocesses are nonetheless predictable.

Some years ago two heterodox evolutionists presented an account of evolution that was not by slow and insensible degrees but a process that came about in unpredictable jumps, a theory they termed "punctuated equilibria." This theory, not to say the men themselves, was attacked by orthodox evolutionists for whom the automaticity of evolutionary process is an unassailable doctrine. One of the proponents of punctuated equilibria, Stephen J. Gould who had become the best known popularizer of evolutionary theory, was attacked because his account was thought to be giving ammunition to religious opponents of evolution.[1] Gould's probabilistic view of evolution was not, however, intended to provide a basis for religious belief. He angrily disavowed that his ideas had such meaning and said they were being misused by religious defamers of evolution. However, neither did Gould relent in his indeterministic view of evolutionary process. He stated that if you played the tape of evolution over again, starting on earth 4.5 billion years ago, the result we now have would not happen again.[2]

Gould's orthodox evolutionary critics, such as Dennett and Maynard-Smith, were acutely aware that a view of evolutionary process that allowed too great a range of unpredictable events made the entire process so far outside the acceptable range of idealized evolutionary explanation that religious critics could assert that evolution was providential and not accidental in nature. Further, two of the elements of chance that were proposed to have interrupted the smooth chain of evolutionary development were said to have come from outer space, appearing at the most consequential times for human evolutionary development. Namely, this happened at the point where non-living matter somehow becomes living tissue and then again at the point required for the ancestors of humanity to evolve. It is as if the famous mural of Michelangelo on the Sistine Chapel depicting the creator God giving life to Adam had a literal, not a metaphorical significance; the finger of God is detectable at precise points in the evolutionary process.

The present situation is that even as orthodox advocates of evolution have made ever-stronger claims on its behalf, these two specific and well-supported scientific hypotheses have interrupted the deterministic account of evolutionary theory and in both cases, the events came from outer space. In the beginning is the transition from nonliving to living entities—that is, from a collection of inorganic chemicals such as nitrogen, water, carbon, and hydrogen to the first primitive but complex and self-active living cell. The difficulty for evolutionary theory is not only the lack of fossil evidence that might document this transition, but that evolution cannot in principle explain the transition from nonliving to living entities by means of its basic explanatory principle of natural selection. Natural selection explains how species evolve by means of competition among living beings to produce successful offspring, but no such contest can take place between a set of inorganic chemicals and a living cell, or between two sets of inorganic chemicals since both would be nonliving, passive entities. Competition cannot explain the origin of life where there is no possibility of a contest.

Evolutionists, including Darwin himself, have attempted to bridge this gap by suggesting that, given the presence of the required chemicals in the early oceans of earth, enough time is available for a series of accidental contacts among the different chemicals to have taken place and produced the first living cell. A significant piece of evidence for this scenario is a famous experiment done by Miller in the 1950's in which liquids and gasses approximating those that appeared on the early earth were contained in a glass vessel through which an electric current was passed. After several days, the inorganic components assembled themselves into chains of amino acids, the "building blocks" of protein molecules which, in turn, are the building blocks of living matter (i.e., cells). In this way, the dramatic first step from nonliving matter to life was taken. Since then however, there seems to be a block in the process. No empirical evidence has appeared and no experiments have been successfully performed which show how the amino acids and other basic organic chemicals assembled themselves into protein molecules—much less living cells. As a result, new theories of the origin of living matter have been put forward, for instance, that life evolved on land in clay, which provided a kind of material template where cells could evolve.

One effect of the lack of evidence for this transition has been to acknowledge that even the approximately one billion years available from the time of start of the earth until the first signs of life does not provide enough time for enough accidents to cause the origins of life. Thus, Crick, co-discoverer of the structure of the DNA molecule, theorized that since there was not enough time for the development of life from inorganic materials that some form of primitive life even if only basic bio-chemicals must have come from outer space.[3] The usual reference in evolutionary presentations to "billions of years" implies that there is an unlimited amount of time for processes that are otherwise improbable to take place—the vast extent of primordial time becomes a cause and not merely a temporal background where evolution takes place. However, the 4.5 billion years of earthly existence is also a limit to how much actual time is available. As Crick analyzed it, since "the last common ancestor" was already on the earth 3.5 to 3.6 million years ago, and the earth is 4.5 billion years old, then there is "only" a billion years for the development of life to have taken place, and that is simply not enough time. There are at least five definite steps to be overcome to arrive from a purely inorganic chemical bath to the first living cell (the so-called "last common ancestor"): (1) from the prohibitively hot and chaotic environment of the newly formed earth to the point where seas have arisen which contain the chemicals required for life; (2) from simple inorganic atoms and molecules to the much more complex ones of basic organic compounds; (3) from the basic organic compounds such as simple amino acids and sugars to the more complex chains of carbon based bio-chemicals, possibly including RNA chains; (4) from the complex organic compounds and structures to the first single cells; (5) reproduction of cells themselves. These steps must be accomplished in a billion years, but if each of the five steps were to require one-half of a billion or one billion years (a total of 2.5 to 5 billion years)—the length of time seemingly required for such steps to have occurred purely by random processes–then there simply is not enough terrestrial time available.

As in a science fiction movie, a possible answer to this dilemma has come literally from outer space—just in time to save the evolutionary time-line. The alien form in real life, however, is not a humanoid come to warn mankind about the dangers of atomic warfare as in *The Day the Earth Stood Still* or aggressive and technologically advanced insects as in the *War of the Worlds*, but rather meteorites whose intervention has been benign and not destructive. Meteors and asteroids that have fallen onto the earth's surface have been the subject of careful study and analysis that has revealed the presence of amino acids and other simple organic compounds which are the basic building blocks of cells. If we hypothesize that amino acid bearing meteorites fell to earth at the beginning of the earth's existence and integrate that into the evolutionary time-line, about a billion years can be taken off. It has even been proposed, albeit without actual examples, that meteorites may have born not merely the basic chemicals of life, but brought living cells to the ancient earth. The idea that life originated not on earth, but in outer space is not a new one. For example, the Swedish chemist Svante Arrhenius had proposed a theory of "panspermia" in 1908, and Polish science fiction novelist Stanislaw Lem in a 1970's short story mordantly pictured life beginning on earth as the result of alien visitors dumping their waste bucket onto the primitive earth![4] But even prior to the scientific hypotheses, the idea that the region of outer space in which the Sun, the stars, and the planets exist is not the cold, dead region pictured by modern science but a kind of cosmic womb is an idea with ancient provenance. Thus, a religious author strongly critical of modern science, C.S. Lewis, pictured the outer space surrounding our solar system as a font of life-giving rays.[5]

If meteors bearing the basic chemical building blocks of life were the first intervention from outer space, the evolution of the human species was aided by a later intervention, namely, the meteor or comet whose collision had the effect of destroying the giant dinosaurs that then roamed the earth. This is the well-known hypothesis that a meteor collided with the earth and left a big hole in its surface, throwing up enough dirt and dust to totally block sunlight for a number of years. This intervention from outer space was catastrophic rather than benign in its immediate effects in contrast to the theory that meteors brought amino acids to earth; however, the hypothesis that a comet killed the dinosaurs has better empirical proof and stands as a nearly confirmed hypothesis.[6]

The account of a meteor descending from outer space splitting the sky with a roar and flash like a monstrous lightening bolt, killing off the terrifying tyrannosaurus rexes, huge brontosauruses, and back-plated stegosaurs, is cinematic to the point of seeming fictional. However, the empirical evidence that such an event actually happened is eminently scientific—while observable, it also requires one or two steps of inference to make the case. The evidence lies primarily in the fact that over all the earth, at approximately the same level below the surface and within the same level of ancient geological strata, there is a very thin but detectable layer of the element iridium. Iridium is a rare element, which because of its weight and atomic composition is rarely separated out in an unmixed form from other elements (this can only be done by immense pressure and heat). A pair of scientists named Alavarez, a father and son, proposed the now widely accepted hypothesis that the iridium layer was the result of an enormous comet hitting the earth and throwing up enough hot detritus to have laid down a layer of iridium that was separated from the other materials it was originally mixed with.

Such a catastrophic event would naturally have blocked the sunlight for at least a year from reaching all parts of the earth. This coincided with a proposal already in the air, namely, that the reason for the death of the dinosaurs was just such an event as the blocking of the Sun's direct rays. The reality of comets is well proven scientifically as there are large lake basins and deep circular depressions over all earth, but the point of a very large comet or asteroid is that by dealing death to the large dinosaurs, it also made possible the evolution of man. The death of the dinosaurs evacuated a large number of ecological niches that mammals could then take over without competition from the giant lizards which had hitherto dominated. Thus, the death of the dinosaurs was one of the necessary conditions for the evolution of mammals, primates (an advanced class of mammals), and the human species—which is the most advanced of all the primates and of all living things on the earth.

The view of the evolution of the human species, which includes two necessary events which arrived unpredictably from outer space, obviously means that the process of human evolution did not take place inevitably once given an initial set of conditions but in fact required adventitious interpositions. The evolutionary ideal of slow development and gradually increasing complexity of species over eons of geologic time now seems impossible to hold. With the automaticity of the evolutionary process now negated, it is as if in the automatic machine-like process, the machine had to be stopped for a moment while an external and intelligent agent made a critical adjustment. Not only are the interruptions from outer space beyond the acceptable range of probability for evolutionary process as usually accepted among orthodox evolutionists, they also allow for the rational supposition that an external agent had interfered with the evolutionary process to produce the desired result, namely, the evolution of the human species. And since we are dealing with a cosmic process, not a machine shop, that agent is more than a human engineer—an agent intelligent and powerful beyond human ken or divine.

Such a theological inference is subject to the accusation that it refers to the so-called "God of the gaps," that is, the impulse for religious believers to interpose divine action at those points where science cannot explain things—explanatory gaps in effect. Thus, for a while religious believers could claim that scientists could only create chemical compounds of the more basic sort, but that creating the chemicals found in living matter was beyond science and the province of the creator God only. This belief was proven false in the early nineteenth century when chemists were able to create organic compounds found in living organisms from inorganic chemicals. This proved that the so-called "gap" between the organic and inorganic realms did not exist and could be overcome by sophisticated laboratory processes.[7] However, in the cases we are considering here, the gaps have not been overcome by laboratory experiment. This is not to say that it will never happen that some simulacrum of a living cell will not be produced by means of a laboratory experiment, or that, *Jurassic Park* style, some usable dinosaur DNA will not become available.

The true gap in the explanation of the origin of life is not that of chemical combination, but that of agency. The chemist Liebig was the agent who "created" organic urea from sulfur, water, etc.; likewise, in the evolutionary timeline, the agent is supposed to be a concatenation of accidental physical events, which in turn requires that there be enough time for the required series of accidents to have taken place. Further, the gaps we are discussing here are gaps indeed, the questions of how life began and of how human beings came into existence. These two "gaps" are really evolutionary chasms because they are beyond the explanatory range of natural selection, and it is the solution of bona fide scientists, who are usually agnostic that the answers come from outer space. Such reliance on comets and meteors hitting the earth heightens the probabilistic aspect of evolutionary process to the point where postmodern man, like his premodern forebears, looks to the heavens for inspiration and the source of life and human existence.

Notes

1. Sagerstrale, *Defenders*, 122–126. See also Daniel Dennett, *Darwin's Dangerous Idea* (New York: Simon and Schuster, 1995), 282–299.
2. Stephen Jay Gould, *Wonderful Life* (New York: Norton, 1989), 292–323.
3. Francis Crick, *Life Itself* (New York: Simon and Schuster, 1981), 17–129.
4. Ibid., 7, 8.
5. C. S. Lewis, *Out of the Silent Planet* (New York: Scribner, 1938), 31.
6. Stephen Jay Gould, "Sex, Drugs, Disasters and the Extinction of Dinosaurs" in *The Flamingo's Smile* (New York: Norton, 1985), 417–426.
7. Steven Weinberg, *Dreams of a Final Theory* (New York: Pantheon Books, 1992), 259.

10

Evolutionary Psychology and the Inevitability of Religious Belief

It is often said that since 9/11 everything has changed" This is not necessarily the case since such sudden, traumatic events often bring to awareness something that has already been in process but not fully recognized. In the case of the attacks on the Pentagon and the Twin Towers, it is the anger of Islam toward the West. This is also the case concerning the science and religion conflict, since it is now impossible (post 9/11) to deny the survival, not to say the growth, of religious belief since the Enlightenment. One of the premises of enlightenment politics and scientific philosophy has been the strong assertion that religion, particularly in its revealed form—Judaism, Christianity and Islam—was doomed to inevitable disappearance in advanced civilizations. Religion, it was said, provided a competing but necessarily deficient account of the universe compared to science.

Religious belief can be defined as that aspect of human nature that searches for meaning in the form of belief in God or gods and in a spiritual realm that underlies or transcends the physical realm. The reductive tendency of science explicitly challenges the rationality of religious belief, reducing religion to a primitive fear reaction in the face of death and unpredictable disasters such as earthquakes, floods, and famines, or to a prescientific means of explaining physical phenomena. In its more aggressive form, the secular scientific tradition has what amounts to a "search-and-destroy" mentality when it comes to religious belief that seeks naturalistic and reductive explanations to fully account for the causes of religious belief.[1] The belief in universal progress that remains part of the secular scientific mind-set assumes that religious belief will eventually be overcome by the more rational methods and explanations of modern science, a point made insistently by Freud, Marx, and Comte among others. Thus, it is of major significance when scientific research comes to conclude that religious belief is an inevitable part of our common human nature, particularly when the specific field of research is evolution in the form of evolutionary biology.

101

The battle between the evolutionists and counter-evolutionists seems to have no end. In this battle, the conflict between science and religion comes to its final and most bitter point, i.e., the application of the mechanical view of nineteenth-century deterministic science to the issues of human origins and human nature. At stake is whether mankind is just another species, an accidental transition point in the evolutionary process, or special because we are not only one of God's creatures but the apex of the Lord's creation. The advocates of both science and religion strain at their utmost since in the evolutionary controversy the whole game is one the line. However, the conflict over evolution has changed over time; we are now in the twenty-first not the nineteenth century, and the grounds of the battle have shifted. The issues that one hears about today are no longer as likely to be an exchange between fundamentalist defenders of the Bible versus evolutionary materialists, but of two new developments: "intelligent design theory" and "evolutionary psychology."

Intelligent design theory is being advocated by a group of believing scientists and mathematicians who focus on the inherent weakness of the theory of evolution. That is. it is just too improbable to believe that the various interdependent organs and biochemical processes within each organism, as well as the origin of each species came about in the machine-like and inevitable manner that Darwinian evolutionism asserts. Such critical observations point to gaps in the evolutionary argument into which the intelligent design theorists insert the creative power of God. In their view, it is not just that evolution lacks certain fossil evidence or a clear understanding of protein chains, but that there are large fissures in the evolutionary account that evolutionary concepts cannot bridge. Scientists, such as Behe, and mathematicians, such as Dembske, have given detailed analysis and comprehensive arguments to make this point in their books, and an organization advocating intelligent design is quite active. However, intelligent design theory is attacked by orthodox evolutionists as bad science for which no evidence properly exists, and they continue to assert that the evolutionary paradigm can explain all the significant issues and answer all objections.[2]

In general, the criticisms of intelligent design theory of evolutionary theory have great strength since much of the argumentation of evolutionary theory is rhetorical and inferential rather than based on direct empirical evidence. The extravagant claim that evolution refutes the idea that human life has meaning and proves that nature is not a designed artifact is not well supported and philosophically weak. But such criticisms are metascientific rather than a proper part of empirical biology, and so it seems that the intelligent design theorists went wrong when they tried to use the law to force their theory as a part of high school biology courses. George W. Bush commented that such criticisms are appropriate in the social studies curriculum (rather than the science curriculum) since the materialism and atheism often associated with evolution is a misuse of science rather than a legitimate part of it. Intelligent design theorists are in a way attempting to co-opt evolutionary theory by combining philosophical criticism and religious doctrine with scientific knowledge, and the advocates of the secular regime are resisting mightily.

If the counter-evolutionists have become more exact and sophisticated in their approach, so also have the evolutionists. No longer asserting blatantly that man is descended from monkeys, contemporary Darwinism has evolved to a point where animal behavior is currently being explained in terms of genetic controls and evolutionary drives. The original subfield, which applied modern genetic and evolutionary theory to animal behavior, was Sociobiology, but more recently other terms have been applied to this field, such as "behavioral ecology" or "evolutionary psychology." Evolutionary psychology is touted as the final victory of evolutionary materialism because it is said to explain human as well as animal nature, and that, in effect, no additional principles are necessary, according to Wilson, Cosmides, Toohy, and other proponents.[3] They claim that evolutionary laws, which apply to ants, turkeys, and gorillas, apply equally to mankind, and those aspects of human nature which seemingly make us special and differentiate our species from all the other animals are completely explainable by evolutionary psychology. Contemporary evolutionary biology has reduced the image of man from "little less than the angels," as in the *Psalms*, to that of "the naked ape."

Interestingly, the main obstacle to complete reduction for the evolutionary psychologists is not intellectual but ethical, that is, they are less impressed by our species' ability to reason than by our ability to make sacrifices on behalf of each other. The reason evolutionary theory is concerned with what they dispassionately term "altruism" is not hard to find; Darwin knew that there are examples of organisms that sacrifice themselves for other members of their species, thus, contradicting the central Darwinian thesis: that the struggle for survival is the biological engine of evolution. The solution to the problem of "altruism" came during the 1970s and 80s when genes and DNA were brought into the theoretical mix of evolution along with population theory. The organism sacrificing itself, such as a mother bird distracting a predator about to attack the chicks in her nest, is acting not to preserve herself but her genes, which are inherently part of the DNA of her offspring. E. O. Wilson, often noted as the founder of sociobiology, explains the problem and its solution this way:

> This brings us to the central theoretical problem of sociobiology: how can altruism, which by definition reduces personal fitness, possibly evolve by natural selection? The answer is kinship: if the genes causing the altruism are shared by two organisms of common descent, and if the altruistic act by one organism increases the joint contribution of these genes to the next generation, the propensity to altruism will spread throughout the gene pool.[4]

With this solution in hand, the gene (selfish and otherwise) and kin relationships became the theoretical means by which sociobiology and its nominal successor evolutionary psychology set out to explain all aspects of animal behavior including human behavior, and especially concentrating on those aspects of human behavior, such as altruism, politics, economics, mutual support, and family ideals, that most clearly distinguish us from the rest of the animals. Not only is genetic science used to explain such phenomena, reference is also made to the findings of social science fields such ethology, anthropology, and animal psychology to establish the pathways from particular genes to exact patterns of behavior which must be discovered by observation of the organism. Since it is based on evolutionary biology and social science, evolutionary psychology is reductive and tends to be materialistic and to exclude idealistic, religious, or philosophic rationales for the behavior of human beings. Reductivist as it is, the agenda of evolutionary psychology has met with strong resistance, particularly from the cultural left. The reason for this is revealing; the characteristics of human nature and of human society highlighted by evolutionary psychology contradict the elevated and idealistic views of human potentiality of the left. Furthermore, evolutionary psychology is built on the assertion that human beings share a common biological structure such that human nature and human society are built in a certain manner which cannot be changed no matter what laws are passed or how much social influence is brought to bear. This is inevitably frustrating the totalitarian and the ameliorist urges of left-wing politics.

Since its particular emphasis is to explain altruistic acts, evolutionary psychology has been applied largely to ethics when considering human behavior. However, evolutionary psychology includes religious belief among those characteristics that are common to all human cultures, part of human nature, and that are fully explainable by evolutionary principles. This is a new scientific approach to religious belief since evolutionary psychology does not denigrate it by stating that religion and ethics (understood as the notion that there are ethical norms that exist objectively) are childish notions to be outgrown by educated and serious people. Rather, it accepts that religion and ethics are as inevitable a part of human nature as sex or self-interest and thereby should be included in the general evolutionary account. In prior accounts from a scientific point of view, religious belief was something to be overcome by our increased knowledge of biology and science, something only the ignorant, fearful, or primitive needed as a security blanket, or as an explanation for phenomena which would now be replaced by scientific explanation. In those days, the origin of the human species was at stake and so the older evolutionary stance opposed revealed religion, and full scale attacks on the Bible were required in the attempt to replace the biblical account of the creation of Adam and Eve in *Genesis* with the new account contained in Darwin's *Descent of Man*. Nowadays, however, religion is regarded as an inevitable and necessary feature of human nature and is sociobiologically understood.

As evolutionary psychology has considered altruistic acts, it has offered such theories as "kin selection" and "reciprocal altruism," which explain self-sacrificial acts by human beings as strategies to preserve and pass on their genes. This kind of reductive explanation, however, gives out at the point when self-sacrificial acts pose no plausible rationale for gene preservation. In a remarkable confrontation, writer Malcolm Muggeridge once posed the example of Mother Teresa to Edward Wilson to make this point.[5] But such extreme examples are most often motivated by the religious belief of the altruistic actors and so religion becomes the final bar to the completion of the reductive agenda of evolutionary psychology. The response of evolutionary theorists to the now undeniable persistence of religious belief has been a premature declaration of victory in the battle to reduce religion to evolutionary tendencies, as by E. O.Wilson, Richard Dawkins, and even Daniel Dennett in his demand that religious belief be outlawed. Wilson writes, "as always before, the mind cannot comprehend the meaning of the collision between irresistible scientific materialism and immovable religious faith ... our schizophrenic societies progress by knowledge but survive on inspiration derived from the very beliefs which that knowledge erodes."[6]

Some current theorists have attempted a fuller explanation of religious belief following the rubrics of evolutionary psychology by integrating data from the behavioral and cognitive scientific fields with evolution. Among the most cited is Pascal Boyer, who despite the fact that his first name is the last name of the great religious philosopher, is an avowed atheist. Boyer's approach is subtle, for he explicitly warns evolutionists and free thinkers not to ascribe religious belief to anything as simple as a "sleep of reason," a prescientific attempt to explain natural phenomena, or a psychological "urge." Rather he says that there is no one common structure underlying religious belief, for it is the result of a combination of several structures which nature uses for other, nonreligious purposes: "religious notions are not at all a separate realm of cognitive activity."[7] Boyer's explanation allows him to acknowledge the inevitability of religious belief while at the same time denigrating it—as if natural selection never really intended it. But Boyer's explanation raises more questions than it answers since it does not provide a sufficient explanation of why religious belief evolved as strongly and persistently as it has if it is merely an evolutionary accident. If religious belief has no selective advantage, why then does it persist? Moreover, even accepting Boyer's account, does this not imply that religious belief is therefore a reflective, second-order aspect of human nature rather than the result of a primitive biological urge? St. Augustine said that the term "religion" derives from the Latin to "re-read" (or "re-lege"), implying that religious belief is something that comes upon reflection or reconsideration. Such a meaning agrees with Boyer's account that religious belief is a second-order, rationally derived characteristic; it is the result of reflection, therefore, and not of impulses born of fear or ignorance.

A cruder approach to explaining religion biologically is simply to assign it a genetic cause, literally a "God gene," which by the production of chemicals in the human brain gives a "high" to a person's mood and so becomes translated into religious belief. The medieval cathedrals, multiple creeds, ancient temples, ethical writings, universality of religion throughout the world, intense human emotions, aspirations to transcendence, transnational social organizations—in short all those aspects which characterize religion at its highest expression are explained by a single genetic mechanism. This approach, so broadly reductive as to be rightly termed simplistic, is nonetheless currently being pursued by some researchers.[8] But however crude and however reductive, such research gives one assurance in the great debate, namely, that religion and religious belief are permanent aspects of human nature, even more so since in this simplistic genetic account they are "hardwired" into the human genome.

The persistence of religious belief remains a major challenge to evolutionary psychology and, according to E. O. Wilson, dealing with religious belief is sociobiology's biggest challenge. "Religion constitutes the

greatest challenge to human sociobiology and its most exciting opportunity to progress as a truly original theoretical discipline."[9] Later evolutionary psychologists who have attempted to explain religious belief
in biological terms have not accomplished it in any satisfactory way,
and the reason is that contradictions emerge when religion is explained
in terms of natural selection. By contradictions, what is meant is not
simply explanatory difficulties that further research can be expected to
overcome, but rather explanatory fissures inherent within the attempt
itself that occur when evolution is applied to religion.

Here are three contradictions evolutionary psychologists face when
attempting to explain the reality of religious belief.

1. Religious belief does have obvious selective advantages in that
typically religions promote large families, which has the effect of transmitting the parents' genes to the next generation. On the other hand,
many religions promote celibacy and dedicated virginity, which is just
as obviously a self-defeating strategy for transmitting one's genes.

2. Religion inculcates a worldview that brings about psychological
benefits since belief in a divinely created world order brings confidence,
dampens anxiety, and makes the believer a more effective person. On
the other hand, religion sometimes makes the believer overscrupulous
in performing religious duties, making him feel an oppressive sense of
guilt and unfilled obligation that suppresses an individual's sense of
freedom and drive.

3. Religion in the sense etymologically as "binding" or "re-tying"
(from the Latin "*re-lige*," a different etymology from St. Augustine's mentioned above) provides strong social identity and often has been the factor
in keeping an ethnic groups whole while undergoing severe oppression,
such as the Mormons or the Poles. On the other hand, religion sometimes
brings about an individual's conviction that their own conscience is the
direct voice of God, as in the instances of Martin Luther and Joan of
Arc, which sets them in direct conflict with social authorities and causes
disruption of the social fabric. These contradictions make it questionable
whether it is possible to render a comprehensive evolutionary explanation
of religious belief and, at this point at least, it has not been done.

Notes

1.   Daniel Dennett, *Breaking the Spell* (New York, Penguin, 2006).
2.   Michael J. Behe, *Darwin's Black Box* (New York, Free Press, 1996); www.
      discovery.org/csc/.
3.   Leda Cosmides and John Toohy, "Evolutionary Psychology: A Primer," Center
      for Evolutionary Psychology, www.psych.ucsb.edu/research/cep/primer.html.

4.   Wilson, *Sociobiology*; 3.
5.   Edward Wilson, *On Human Nature* (Cambridge, MA.: Harvard University Press, 1978), 165.
6.   Ibid., 172.
7.   Pascal Boyer in *Skeptical Inquirer* magazine; www.SCICOP.org/si/2004-03/religion.html. Also, see Boyer's *Religion Explained* (New York: Basic), 2001.
8.   The cover story in *Time* magazine, "The God Gene," (October 25, 2004), features the research of Dean Hamer and his book, *The God Gene: How Faith is Hardwired into Our Genes* (Garden City, NJ: Doubleday, 2004).
9.   Wilson, *Human Nature*; 175.

11

The Agony of J. Robert Oppenheimer

"Oppie," as he was known to his friends, is a contradictory and foreboding figure in the culture of science. He is, more than Einstein, Newton, Darwin, or Galileo, representative of both the high intellectuality and the tremendous power of modern science. This intellectuality is so abstract that there is no hope of truly understanding science without a firm grasp of the higher mathematics, and this is a power that gives governments means to destroy not only the cities and armies of their enemies, but all life itself. Known as the "father of the A-bomb," a title he cherished yet which he found laid on him a tremendous sense of guilt, important in the development of twentieth-century physics during a time of revolutionary discoveries, he is important less for his own discoveries than for importing quantum mechanics to America by forming a school of theoretical physics at Berkeley in the 1930's. Condemned after World War II for his intimate connections with members of the Communist party and his support for left-wing causes during the 1930's, he used all of his energy and talents to maintain his position as a powerful counselor to presidents and secretaries of state and war on atomic matters during the breakout of the Cold War with Soviet Russia. Like Galileo, Oppenheimer was subject to a prosecutorial hearing for which he is revered as a martyr to the freedom of scientific thought, yet his previous lying and contempt for others sometimes expressed in outright cruelty make him an unsympathetic character. Although a prominent physicist who dealt with Einstein and Bohr as equals, his unparalleled influence after World War II was less due to his scientific accomplishment than to his ability to use language in a uniquely persuasive and outright poetic style. Finally, as a representative figure of modern science, he compels attention not because he invented an important new theory but because of the appearance that he gave of unresolved personal struggle; he was a scientist with *angst* and, thus, a fitting representative of the atomic age.

The question of Oppenheimer's membership in the Communist party remains a hot topic among historians and researchers, but in the end, the question seems unresolvable.[1] The question is complicated because

Oppenheimer was admittedly a "fellow traveler" at the time. He attended party functions, gave serious amounts of money directly to people who he knew were members of the party, and actively participated in a number of causes sponsored by the Communist party. The most telling evidence of his close connection to the party was personal: Oppenheimer's wife Kitty, his former fiancée Jean, his brother Frank, Frank's wife, and his best friend on the Berkeley campus, Haakon Chavelier, had been, in the old phrase, "card carrying members of the Communist party." As for Oppenheimer himself, he denied under oath several times that he had ever been a member of the Communist Party.

The fact is that Oppenheimer's documented connections with the Communist Party were so close and intimate that the question of his official membership is a moot point; this was the opinion of an Army security official who was one of Oppenheimer's severest critics. It is not, however, a trivial debate since after World War II he was advocating the internationalization of atomic power and materials such as Uranium, which is necessary to making bombs. Oppenheimer was also accused of doing everything he could to slow down the development of the apocalyptic H-bomb at a time when the Soviets had already embarked on a program to develop this "nuclear weapon."[2] However, no one before, after, or during the notorious security hearing which deprived him of his official security clearance ever claimed that Oppenheimer was disloyal to America or had given away its atomic secrets. "Dammit, I happen to love this country," he once told George Kennan in his overblown fashion.

In a literal sense, it was true that Oppenheimer loved America; he was fond of the landscape of the Rockies and had a home built in the uplands of New Mexico where his father had sent him during his teen years to toughen him up and get him away from his smothering mother. Oppenheimer had been sent to high school at the Ethical Culture School in New York City where he was a precocious, thin, and socially awkward teenager in need of developing his self-confidence to a level where he could enter Harvard. As a boy, he had turned his rock collecting from a hobby to a professional level study of mineralogy; and at Harvard, he graduated as a chemistry major in three years while taking extra courses in literature and physics. He went on to study in Europe and, after a false start as an experimental physicist, found his métier in the pursuit of theoretical physics, which was then in a phase of enormous development not seen since the seventeenth and eighteenth centuries—when the "new physics" of mass, forces, and calculus was invented. After attaining his doctorate in quantum mechanics and publishing a series of original research papers, Oppenheimer returned to America, where his reputation had preceded him, to teach concurrently at the California Institute of Technology and the University of California at Berkeley. At Berkeley, Oppenheimer established an outpost of the new theoretical physics, training a remarkable group of graduate students; at the same time,growing into his own shape as a learned professor on the Berkeley campus, socializing with his peers and becoming involved with the radical and leftist politics current at the time.

The course of what might have been an outstanding but predictable professorial career was changed fatefully by the confluence of two events: World War II and the splitting of the atom. Oppenheimer was inescapably attuned to both events as a leftist who had sent funds to support the Republican, anti-Fascist side during the Spanish Civil War and as a theoretical physicist who knew of atomic research firsthand and was directly involved with it. Ernest O. Lawrence, who was a colleague at Berkeley, had invented the *cyclotron*—the first particle accelerator based on the experimental fact that the atom was not an indivisible solid mass—but a congeries of particles separated into a nucleus and outrider rings of electrons. German physicists had split the Uranium atom into two smaller atoms of Beryllium, a process that at the same time released two heavy particles, or *neutrons*, which could then split other Uranium atoms producing the famous chain reaction. The result, as predicted by scientists, would be a massive explosion of then unknown force, but surely well in excess of that produced by chemical explosives such as TNT. The study of the interior of the atom and the shape of the physical universe, which was the aim of quantum physics and relativity theory in the early twentieth century had produced not just the greatest intellectual achievement of all of modern science, but also the means of constructing the most destructive weapons imaginable.

With knowledge came power and with power came responsibility—a truth from which Oppenheimer never shrank. With his graduate students he did early research into the details of the projected A-bomb, including how much fissionable Uranium would be needed and how big the explosion would be. He attended conferences with physicists who were also studying the feasibility of nuclear explosive devices (including the H-bomb) where his evident knowledge propelled Oppenheimer into the forefront. Although there was much fitful politicking involved in the selection process, he was eventually chosen by General Leslie Groves to head the research effort to develop an American A-bomb before the Germans. Surprisingly to many people who knew him well, Oppenheimer was an excellent choice to head up the A-bomb project. While in charge of the research part of the effort he showed himself to be an effective administrator who led his team of world-class scientists and technicians to completion of the A-bomb project in time to drop bombs on two Japanese cities and force a Japanese surrender.

Oppenheimer's administrative ability was of a special kind, for he was an effective leader as he had shown by his work with scientists, high-power intellectuals, and graduate students at Berkeley as well as when he was at Los Alamos during the A-bomb effort and again after the war as head of the Institute for Advanced Studies. He was not an administrator of the mover and shaker kind as was General Groves, who only a day after accepting the government's offer to head up the Manhattan Project ordered the acquisition of ten thousand acres of land in Tennessee for the building of a Uranium processing plant. Oppenheimer's style was rather to talk to small groups of scientists and persuade them by the force of his intellect and with his considerable expressive ability to work toward a common end. Of his intellectual ability as a scientist there could be no doubt, especially as he had not limited himself or his students to one specialized field within physics but to whatever field or development was "hot" at the time. Thus, he was able to show a specialized knowledge in their own field to almost every scientist he met. Further, Oppenheimer had a very wide array of knowledge outside of science especially in literature (he read several languages) but also in art and history. This wide range of interests not only indicated that Oppenheimer was an excellent conversationalist, for a wide array of knowledge and interests is also a sign that an individual has the range of mind which considers the outcomes and alternatives of actions—a prerequisite for outstanding leadership, which, in fact, Oppenheimer provided.

His greatest accomplishment as an administrator was to convince so many able scientists, some of them Nobel Prize winners (including Europeans who had already made their scientific reputations like Hans Bethe and younger scientists who would make theirs later on like Richard Feynman), to come to an isolated encampment in the New Mexico Rockies to work together to make the atomic bomb. Once there, he worked to provide work areas, laboratories, and housing for the scientists and their families. He acted as liaison and buffer between the independent minded scientists and the U.S. Army officers in order to organize work groups to conduct research in separate areas including explosives, atomic theory, and mechanical testing. Most importantly, he worked to co-ordinate the whole to make it all come out right with the production of two types of A-bombs, the Uranium "Little Boy" and the Plutonium "Fat Man" versions, in time to help end the war.

Oppenheimer's accomplishment is historically important and utterly unique. Yet the best known history of the A-bomb, Richard Rhodes' *The Making of the Atomic Bomb*, spends less time on Oppenheimer's administrative ability or his place in the development of twentieth-century physics than on his personal psychology.[3] It is true that Oppenheimer's personality is the most interesting and compelling part of him. He was extraordinarily ambitious and by the end of the war had made an enormous accomplishment that earned him universal recognition, yet he never seemed to be happy or settled in his own skin. Always there was a visible element of self-doubt, or awkwardness amongst people outside his circle of family and close friends. Yet this aspect that made him seem unfinished and which often drew sympathy was complemented by demonstrations of arrogance and self-importance which occasionally yielded acts of outright cruelty when facing opposition. The prodigiously intelligent youngster who was invited to give a lecture to a mineralogical society and the young physicist who called Einstein "cuckoo" for his rejection of quantum theory[4] was always able to fall back on the superiority of his own intelligence when dealing with political enemies such as Lewis Strauss and the U.S. Army security apparatus. He always assumed that he could put such enemies out of play by sweeping them off the field and if necessary destroying them—like Arjuna on the field of battle, the philosophical warrior in the Hindu epic, the *Bhagavad-Gita* to which Oppenheimer often referred.

The *Bhagavad-Gita* was an important part of Oppenheimer's reading, and it is necessary in order to understand him at all to realize how important the reading of books and poetry was to him. Through his reading, he acquired his personal philosophy and ethical beliefs, attempted to understand the intricacies and troubles of his own personality, and acquired his strikingly mannered style of expressing himself in writing and in speech. He read the poets Baudelaire, Herbert, and Donne, and on a walking trip to the island of Sardinia as a young man he overcame a terrible depression while reading the elaborately memorial novels of Proust. As a man aware of his own intellectuality, he searched out the most *avant garde* literary texts; he came, so he said, to his understanding of communism by reading *Das Kapital* on a cross-country train ride, and he learned Italian in order to read Dante's *Divina Commedia*. It seems a natural consequence, then, that he would read the Indian epic, the *Bhagavad-Gita*, and even take lessons in Sanskrit to read it in the original.

This reading and what he attempted to derive from it personally eventuated in the development of Oppenheimer's formal style, which was mannered and has been described as "baroque" and "priestly."

Oppenheimer would rely on this style when addressing colleagues on matters of importance or when making public statements. Upon observing the first explosion of an atom bomb in the New Mexican desert, he claimed that he quoted the Bhagavad-Gita: "Now I am become Death, the Destroyer of Worlds."[5] Upon receiving the Fermi Award from the government, in part as an expiation for the damage done to him during the security hearing in the prior administration, Oppenheimer said, "we are engaged in this great enterprise of our time, testing whether men can both preserve and enlarge life, liberty and the pursuit of happiness, and live without war as the great arbiter of history." Addressing President Johnson who presented the award just weeks after the death of President Kennedy, Oppenheimer continued, "I think it is just possible, Mr. President, that it has taken some charity and some courage for you to make this award today. That would seem to me a good augury for all our futures."[6] He thus raised his personal situation to a historical plane with echoes of the Declaration of Independence and the Gettysburg Address, and then when addressing Johnson used his most sumptuous phraseology "*some* charity and *some* courage." The adjective implied a nicety as if the speaker did not want to embarrass the giver of the award and elevate his own status by stating plainly that it required *great* charity and *great* courage to make it. In fact, it required neither charity nor courage for Johnson, the most calculating of politicians, to make an award that had already been promised to the recipient by Johnson's martyred predecessor. Granted that this event was a vindication for Oppenheimer, his speech implied that more than a merited and delayed recognition of his great service had taken place, extending to self-exaltation.

The style seemingly had two purposes: as a means of manifesting Oppenheimer's own uniqueness and superiority of learning and, perhaps unintentionally, a way of obscuring his real meaning. Oppenheimer may not have intended to be deceptive or evasive, yet his style gave him the means to be so if he desired, and many thought that he did so intend. In any case, his style tended to obscure plain meaning, but to the degree that the style represents the person, Oppenheimer's style implied a personality that had difficulty in arranging things such that he could be at ease with himself and clear about his own motives and desires. He had seen two psychiatrists as a young man after he left Harvard for study in Europe, but as a middle-aged man all of his considerable life-experience and overwhelming accomplishment were not enough for him to able to define himself. Thus, the style was not artificial in this sense but a true reflection of the man Oppenheimer was.

How can the unevenness and self-contradictory nature of Oppenheimer as a person be explained? Isador Rabi, a fellow physicist of high accomplishment who was his life-long friend and defender, spoke of this dissociation. Rabi said that his dissociation was typical of men whose fathers had left the Orthodox Jewish faith, leaving such men without the grounding of moral sense imparted by the traditional faith.[7] By contrast, Rabi's own father remained Orthodox even though Rabi left off being observant. When Oppenheimer invited Rabi to join the Los Alamos group to help build the A-bomb, Rabi refused. In support of Rabi's thesis, Oppenheimer was sent as a boy not to the public high school but to the Ethical Culture School in New York City where Oppenheimer's family lived in well-to-do circumstances. Here, Oppenheimer learned not the *Torah* or other sources of traditional Jewish learning, but an ethics of generalized ethical concern based on progressive sensibilities attuned to social issues of income inequality, the condition of the working poor, racial oppression, and international peace, and which disavowed the tenets of any particular religion.

On the other hand, contradicting Rabi's explanation, Richard Feynman and Isaac Asimov also grew up in New York City about ten years after Oppenheimer did, albeit in poorer circumstances, and their fathers as new immigrants had left the Orthodox faith just as Oppenheimer's father had. However, there is no evidence of a disassociated personality in either of their cases as attested by reports of them by friends and family and by their autobiographies.[8] Both Feynman and Asimov had famously healthy egos and integrated personalities, as did Oppenheimer's brother, Frank. (Interestingly, Asimov became irate with his father when late in life the old man returned to the practice of the Jewish faith.) Seeking another explanation of Oppenheimer's disassociated personality, we could look to the earliest influences on him including his preciosity, devotion to reading and intellectual pursuits, his shyness and social awkwardness, and his mother's over-protectiveness. However, psychoanalysis based on early influences is uncertain, often leading the observer to make imaginary inferences rather than rendering reliable explanation, and so finally, we can look to the obvious outward manifestations of Oppenheimer's personality where we can reasonably infer two dominant tendencies, namely, ambition and guilt.

Oppenheimer's ambition was manifested in the self-dramatizing in his language and even in his posture and was apparent in his immediate and unreflective acceptance of Grove's offer to lead in the development of the A-bomb. The ambition is even more apparent after the war when he positioned himself as the chief liaison between the scientists who had developed the A-bomb and the politicians and generals who were making the policy decisions. "Oppie" was well set to do this because of his reputation as "father of the A-bomb" and as head of the influential General Advisory Committee, which advised the early Atomic Energy Commission. His guilt, reflective of the common guilt felt by the scientific community after the bombs were dropped on Hiroshima and Nagasaki, probably led him to believe that the sharing of the secrets of atomic power both for warfare and peaceful purposes was in the best interests of the future of the human race. Guilt may also have led him to a negligent attitude about security at Los Alamos, and guilt may have been the spur that led Oppenheimer to oppose the expansion of the explosive power of atomic weapons and suppress the development of the H-bomb.

The great calamity of Oppenheimer's life was the Gray Board hearing called by his political opponent Lewis Strauss, then head of the Atomic Energy Commission, to decide whether Oppenheimer was a security risk. The Cold War was heating up as Stalin conspired to put the nations of Eastern Europe under Soviet domination and China fell to the communist insurgency; the Soviets exploded an A-bomb of their own and were working on developing an H-bomb. In this context, Soviet Russia had turned from a noble ally in the war against Hitler to a mortal enemy of the United States, and, unfortunately for Oppenheimer, his tight connections with communist causes and colleagues in the 1930's was no longer seen as a sign of his social idealism but an indicator of possible traitorous liaisons. Oppenheimer, however, had recognized in the late 1930's from published reports of the infamous purge trials and from the on-site testimony of the false imprisonment of scientists that the Soviet regime was morally corrupt. The charge against Oppenheimer was not that he was a Soviet spy but that he was a "security risk," and the hearing itself was not a criminal prosecution but merely an administrative hearing. Both of these facts, however, hurt Oppenheimer's chances of clearing himself of the charge as the AEC didn't have to follow the stricter protocols of revealing evidence to the accused that apply in criminal prosecutions. Also, being an accused "risk," rather than a felon, allowed much greater latitude of interpretation beyond the confines of the criminal law wherein charges and offenses are strictly defined. The case against Oppenheimer was presented by an experienced criminal prosecutor who caught Oppenheimer in contradictory explanations of his past behavior, and who wove his resistance to the development of the American H-bomb into a pattern of suspicious instability. The board's vote was two-to-one against, and Oppenheimer never recovered. He lived the rest of his life as with a mortal wound.

Oppenheimer's trial has been compared to the trial of Galileo. Both men, it is said, made a case on behalf of a new science that their respective authorities found heretical and dangerous to the well being of society. Both men were subject to the horrors of a trial in which they were unfairly and cruelly treated. Yet there are revealing differences; Galileo was fighting for the truth of a scientific idea—Copernicanism—while Oppenheimer was fighting for a policy of dealing with the threat of nuclear war by self-restraint and international controls. Also, the consequences were much harsher for Galileo, who was put under close house arrest and forbidden to publish, while Oppenheimer kept his position as head of the Institute for Advanced Studies and retained the normal freedom of an American citizen. Galileo, however, returned to his early work in mechanics and, despite the restrictions on him, was able to publish his most important scientific work (including the law of falling bodies) in a series of short dialogs that were smuggled out and published in his lifetime. But science advanced much more rapidly in the twentieth than in the seventeenth century, and Oppenheimer would have had to catch up (but never did) if he were to return to active research after a long period of administration. Instead he gave invited lectures and speeches about the atomic age—lectures whose obscurity he himself remarked on.

There is a further link between Galileo and Oppenheimer, one that could seem arcane but one with compelling practical consequences, namely, their involvement with the atom. For Galileo, atomism was a required part of his philosophy of science, for physics could uncover ultimate physical reality only if there were substances whose qualities could be supposed definite and real beyond the errors of sense data and which could be described in a determinate mathematics.[9] But by Oppenheimer's time, the atom had been split and revealed as a congeries of inner and outer particles whose location and velocity could not be empirically measured beyond a certain point and which eluded deterministic description. Over four hundred years modern science had broken down the philosophic atom by making it a subject of intense empirical inquiry. In the twentieth century science had reached its downward limit and had advanced to provide mankind the greatest secrets of the physical universe giving a depth of understanding not available to the ancient Greeks or from the

Bible or any other culture, fulfilling its promise at the most elemental level. But by the same discoveries, science has inescapably given the human race the ability to destroy itself and make the planet earth a place of death and radioactive waste. With scientific knowledge, came power and with atomic power, came a responsibility mankind does not even now seem ready for. In the enigmatic, conflicted, proud, super-intelligent, agonized, and expressive J. Robert Oppenheimer, science has also given us a human representative of the anxiety we all now share.

Notes

1.   Priscilla J. McMillan, *The Ruin of J. Robert Oppenheimer and the Birth of the Modern Arms Race* (New York: Viking, 2005). See also Gregg Harkin and Daniel J. Kevles, "The Oppenheimer Case: An Exchange" *New York Review of Books* 51, 5 (March 25, 2004).
2.   Richard Lourie, *Sakharov: A Biography* (Hanover, NH: Brandeis University Press, 2002), 92.
3.   Richard Rhodes, *The Making of the Atomic Bomb* (New York: Simon and Schuster, 1995).
4.   Kai Bird and Martin J. Sherwin, *American Prometheus* (New York: Alfred A. Knopf, 2005), 64.
5.   Ibid., 577.
6.   Ibid., 576.
7.   Rebecca Larsen, *Oppenheimer and the Atomic Bomb* (New York: Franklin Watts, 1988), 168.
8.   Isaac Asimov, *It's Been a Good Life,* edited by Janet Jeppson Asimov (Amherst, NY: Prometheus, 2002). See also Richard Feynman, *Surely You're Joking, Mr. Feynman* (New York: Norton, 1997).
9.   Burtt, *The Metaphysical Foundations*, 86–90.

12

Scientific Technology is the Return of Magic

Arthur C. Clarke once stated, "Any sufficiently advanced technology is indistinguishable from magic."[1] Surely no one is better qualified to make such a judgment than Clarke who is both a farsighted engineer and a prominent science fiction author, however, while the statement is provocative, it also seems contradictory. On one hand, technology, as the application of modern science to human affairs, implies the debunking of magic. Historically, as we are aware, science replaced magic as a means of explaining and controlling the universe around us. On the other hand, the advanced technology of today does seem magical in its ability to perform astounding tricks such that merely to want something—a new electronic device or a safer automobile, a faster computer or a way to prevent the spread of AIDS—is to have it. It is worth pursuing Clarke's comment first, to understand the historical relationships between magic and religion, as well as magic and science, and second, to understand the manner and consequences of technology bringing about the return of magic in the postmodern world.

Religion, Science, and Magic

As if to show that Clarke's point is not merely rhetorical, there is a significant social movement involving probably millions of people in the Western world which supports the restoration of paganism. "Neo-paganism," as it is sometimes called, is opposed to the revealed religions, especially Christianity. In their desire to return to a pre-Christian sensibility, the neopagans not only worship pagan gods but also have reintroduced the practice of *magic*, including enchantments and spells. In this manner, they have asserted a systematic defiance against not only Christianity, but also science. Magic has made a reappearance in popular culture where it is seen as largely benign-as in A. K. Rawling's stories or the witches on television programs portrayed by wholesome young women. But these examples are the bland residue of a much more seemingly potent reality—that of witches who consorted with demons, sought the destruction of souls, and who were hunted down and killed. In addition, there were satanic cults that challenged the rule of the church by mocking the mass and the ancient worship of gods, demons, and forces that represented nature in its darker and more primitive aspects. Magic is also represented by astrology and alchemy, which stand in opposition to the modern sciences of astronomy and chemistry, although lately there is a tendency to look upon astrology and alchemy as proto-sciences rather than nonsciences.[2] That is, recent historians note not the opposition but the historical transition from magic to science as if the prior magical modes of explanation merely needed to be trimmed away. This left only the factual evidences of the empirical reality rather than the occult powers that supposedly underlie the phenomena—much like how Newton used force and the law of gravity to explain the course of the planets rather than mythological affinities connecting Mars with war and Venus with love. But then Newton, while not an astrologer, did practice alchemy.

One of the first refutations of astrology is contained not in a scientific work, but in the *Confessions* of St. Augustine. Augustine, the fifth-century bishop, was not converted to the Christian religion until he was in his thirties but had such a restless desire to know ultimate truth that he studied a number of approaches current in the days of late antiquity just before the fall of Rome. He rejected the magic of self-professed magicians who would slaughter an animal for a fee to propitiate the gods and promote his own career, saying that he would not even want a fly killed on his behalf, but he was greatly attracted to astrology.[3] As he became more attracted to the Christian faith, however, he developed doubts about astrology; his rejection of magic is characteristic of the revealed tradition in general and stems directly from the First Commandment. "I am the Lord thy God ... Thou shalt have no other gods before me." (*Exodus 20*: 2, 3) The Lord thus announces not only his dominance but his exclusiveness as Israel's only deity, for He is "a jealous God." He prohibits the making of any idols representing anything in the sky, the earth, and the sea and forbids that the Israelites "bow down before them or worship them." (*Exodus 20, 4-5*) The rejection of idolatry subsequently forces the realization that to worship, revere, or to impute a life of some sort to physical and immaterial entities is also forbidden. What is at stake ultimately is the sole reality of the Lord God as the creator of the universe and the father of souls; a doctrine that ruthlessly eliminates all gods, spirits, animisms, and occult forces. (Which is not to say that the ancient Israelites or medieval Christian did not often dally in magic, but such movements were usually harshly proscribed by religious authorities.) We usually take the First Commandment to simply forbid the making of idols, which we understand to be "graven images," such as statues of Diana or of some animistic god. We do not often extend the commandment to apply to such things as a political party, a particular race or class, the free market, or social progress—which are the modern equivalents of graven images. The dominance of the figure of the Lord in the history of Israel has the effect of eliminating not only other gods but magic as well.

If revelation opposed magic from its beginning, the same is not true of the scientific tradition. This assertion runs contrary to the widely accepted view that sees magic and religion as closely related—as if religion evolved from magic and still retains magical elements. By contrast, in this view, science began by closing the door on magical explanations, substituting the vaunted scientific method. However, the actual historical events show otherwise. Kepler and Galileo were not above casting horoscopes thus demonstrating the slow evolution from magic to true modern science. It is well known that Newton was deeply interested in alchemy; he possessed many books on the subject and had done research in it, searching for universal secrets he apparently had not thought he had uncovered in his mathematical physics. Newton's search for true knowledge led him down all the pathways of knowledge of his time, including mathematical physics, alchemy (which concerned the transmutation of metals and changes of substances), and biblical studies by all of which he sought to construct a universal history. His strong interest in alchemy is indicated by the number of alchemical and newly discovered records of alchemical experiments he performed.[4] There is no record of Newton's making any new discovery or offering any mathematical theory in alchemy despite the evidence that he studied the field intently and performed alchemical experiments.

Newton accomplished little in his studies in alchemy but the difficulty arose because of the nature of the field itself; the phenomena were simply too diverse for a unification of the sort that Lavoisier was able to accomplish for chemistry a century later. Alchemy purported to see connections between, for example, the planet Mercury, the element mercury, the mercurial nature of some personality types, the constellations through which the planet Mercury traveled, and the ancient god Mercury including the general aspect of speed, commerce, and communication associated with the god. Alchemical magic combined these disparate aspects with ancient myths from Greece, Rome, and Egypt where Mercury was given the name Hermes to concoct a tradition of secret magic (hence the "hermetic" teachings of the magicians). Believers in alchemy thought that they could use these associations to improve their personal lives in the areas of business and love, for personal control of one's own destiny by means of secret and ancient lore was always at the heart of the appeal of magic and still is if the web sites for magic, alchemy, and astrology are any indication. The connections were so vague as to elude the mathematical treatment, which was Newton's special forte, because the connections were based on affinities and sympathy, in effect *feelings*, which could conjoin all the parts of experience without rigorous description of any particular part. The scientifically useful aspect of alchemy lay in the chemical processes discovered by which alchemists had, for example, combined the element mercury with other elements to form a number of useful compounds. The explicitly magical aspect of alchemy lay in its promise that its practitioners could accomplish the transformation of matter and manipulate events for their own personal use and fulfill their own desires and needs.

The separation of the dross of magic from the gold of science was accomplished over time by means of the discovery of scientific laws and a concept of nature which excluded sympathy and affinity as a method of discovery, substituting hypothesis and experiment instead. As a result, the means of understanding the universe became public, rational, controllable, and the work of experts; it was no longer personal and secret. Modern science forces on its practitioners a kind of intellectual discipline that is missing from the practice of magic. It is true that the scientist becomes much like a wizard, since both have acquired knowledge at a great cost and both have knowledge that must be reverenced because it gives access to secrets which can change matter and influence events to fit our own ends. However, the secrets of science are not really secret and are not deliberately hidden although their inaccessibility is sometimes utilized for commercial or military ends. Access to scientific knowledge is limited by intellectual ability and willingness to study, but it is not a secret cabal as evidenced by the many people who explain science to a popular audience including scientists such as Michael Faraday and Albert Einstein, television personalities such as "Mr. Wizard" and the "Science Guy," and writers such as Stephen J. Gould and David Green. Further, science is presented as important because of its social effects but more important because it is interesting in itself; its descriptions and explanations of physical phenomena intellectually compelling.

By the middle of the nineteenth century, science had developed into the truth teller of Western culture that was independent of magic as well as religion and was engaged in a kind of cultural warfare with them both. However, religion still maintains its categorical opposition to the magical view no less firmly than the secular tradition of modern science;

this opposition is based on the revealed doctrine that the Lord God is the creator of all things "visible and invisible," as the Nicene Creed says. By now it should be possible to tease out the major difference between magic, on one hand, and religion and science, on the other. Not incidentally, this indicates an essential resemblance between religion and science, for the worldviews of both are characterized by regularity, logicality, and laws by which all the various parts of the universe are organized. Further, revealed and secular scientific knowledge describe the universe as an objective system whose primary laws cannot be adjusted or manipulated by human intervention. Divine laws requiring charity and justice impose obligations on believers whether the believers wish it or not, while physical laws describing gravity and entropy are unchangeable by human desire. Stealing and adultery are wrong no matter how intense our desires and we cannot fly into space unaided, nor can we avoid getting old.

Despite forthright opposition from both religion and science, magic is making a comeback in contemporary society, which is apparent in a whole range of phenomena including the increased popularity of astrology, crystal therapies, the occult, witchcraft, and pop psychologies which verge on the magical. In more general terms, the renewal of interest in magic comes about indirectly, not by the revival of ancient heresies and superstitions, but as a result of the influence of the new range and intensity of contemporary technology. Technology has replaced magic because those things we wish for are provided by technical applications of scientific theory to the practical needs of common living. There is now a ubiquity of technological means of providing not only for our needs, but our wants; not only necessities or to relieve hard and tedious labor, but to make our lives easy and to provide interest and pleasure. An example is at hand as this paragraph is written. It is a promotional mailing from a cable company that makes humorous but seriously meant promises to its customers: "MAN STOPS TIME" is the headline for one article that goes on to explain, "Digital Video Recorder ... allows man to suspend time while watching TV." Another piece is headlined, "Exclusive: Man's Sight Improved: High-Definition TV ... provides perfect vision." Just as Alladin got three wishes from the djinn hidden in the magic lamp, so also manufacturers and corporations make every effort to provide consumers with the means of making their lives easier at every possible contingency. Supermarkets and retail outlets show such a diversity and range of options for the consumer as to give a feeling of control and a sense of entitlement so that the consumer becomes indignant when an option is not provided for. The fulfillment of needs, wants, and wishes is assumed as a matter of right since the possibility always seems to exist that technology can supply whatever is needed, wanted, or wished for.

Technology as Magic

The new magic comes without the magical doctrines associated with astrology and alchemy. It is not necessary to know about the god Mercury or to discover the ancient symbolisms that connected all the aspects of Hermetic magic in order to cook a ready-to-serve meal in a microwave oven and, at the same time, to watch a soccer match in Malaysia while sitting in a living room in Boston. There is no witchcraft associated with the technological magic, no dark side, or occult agencies to be propitiated, and thus indulging in magic today does not mean engaging in heresy or blaspheming the name of the Lord. The new technological magic does mean, however, that the recipient of its gifts can ignore the First Commandment, which proscribes idolatry, while staring intently for long periods of time at the images on a computer or television screen. Reverence is thus given to a human made contraption, and the flat screen of the high definition digital television set with a remote control whose signal is received from outer space bounced off a communications satellite (a technology first envisioned by Arthur Clarke) provides the images that hide the face of God from us. It is idolatry for the twenty-first century. However, technological magic hides not only the evidence of God's existence, it also hides the laws and discoveries of modern science. The beneficiaries of technological magic need know nothing about the electromagnetic spectrum, Maxwell's equations, electrical conductivity, the second law of thermodynamics, or the atomic theory of heat and need not ever have conducted a science experiment in order to enjoy barbeque chicken wings hot from the microwave while watching a sporting event half a world away. The new magic requires only the manifestation of desire which current technology will fulfill.

Current levels of technology are different from any that have been seen before in human history. In the early twenty-first century, technology has developed to the point where to understand it as the mere addition of technologies developed over the last century is not enough. Technology develops as evidenced by moving from the telegraph to the to the automobile, to the airplane, to radio, to movies, to radar, to television, to the mainframe computer, to the personal computer, to personal electronic communication and not forgetting the refrigerator, vacuum cleaner, electric oven, microwave oven, freezer, and drip-dry fabrics or the technologies of medicine or of warfare. All this does not give us an understanding of the effect of current technology in general unless one additional premise is thrown in: that enough quantitative changes eventually add up to a qualitative change. After all, it can be argued that technology really begins with the use of stone tools by proto-humans over a million years ago, for human ingenuity combined with the desire to improve our lives has been working throughout human history like a constant engine. Thus, it can be argued that technology has always been a primary force in human history and that human beings have always adjusted to new technologies. Fear of the new is not new, and the addition of new technologies, no matter how radical or at how fast a pace, will not essentially affect the human condition.

In response, the new level of technological application provides an artificial environment that replaces the natural human environment and makes it possible to never experience, much less appreciate the natural environment in which we live our lives. By the word "environment" I do not have in mind simply that living within our cities means not often seeing copses of trees and fields of grass, but the total human environment in which we live. Our suburbs, with landscaped lawns and trees trucked in from the garden center, is a technologized environment that is replanted over the old, natural one. It may have been a hillside on a farm but now has been graded into level lawns and subdivided into one-half acre plots under which water pipes, sewage drains, gas lines, and electrical cables have been carefully planted. Within the houses, a new world of sights from satellites or the cable company and sounds from the attached amplifiers make news, entertainment, and music inescapable—as efficient a method of stifling thought and meditation as could have been deliberately devised.[5]

The social correlation to the need for advanced science to develop theories prior to the invention of computer technologies is that while computers are built as part of an assembly line process, like Model T cars, the real "workers" of the computer industry are usually very bright people (650 or better on their Math SATs). The tinkerers of the computer field are not barely educated geniuses like Thomas Edison, but over-educated geniuses like Steven Jobs. The home and schoolboy inventors are no longer in their basements and garages making crystal radios or television tubes; in today's computerized technology, they are in their bedrooms sitting in front of their PCs hacking into military software and designing computer games. In short, computer technology is a field on which only the very intelligent can play and only the mathematically intelligent at that. It is true that we can open the processor unit, the box that contains the innards, and observe identifiable items such as the fan, the transformer, wires, boards, and chips, and we can even make repairs to some of these items. But the alteration of electronic impulses once they leave the visible wires and enter the chips and boards is really beyond us, for they contain the memory and the calculating capacity whose inner workings are a mystery unless, that is, we happen to be computer engineers or physicists. This is unlike a typewriter or an automobile or steam engine whose inner workings can be understood and often repaired by the average person. The new technology owns us, we are beholden to it because we cannot understand it. Author and critic Bryan Appleyard describes the personal computer as a "black box" and our relationship to it trenchantly.

> At this point the machine becomes an indecipherable black box which, in addition, is likely to be cheap enough to be replaced rather than repaired. Its mechanism is something we have neither need nor competence to explore—the machine becomes as irreducible, as absolute as a natural object ... or as ourselves. It is no longer something to which we have to relate in the way we may have done in the past to cars. Rather it is like a rock or a plant, a part of the natural environment which we pick up in passing and discard when it has expired. It has no interior with which we need to concern ourselves.[6]

The sad and inevitable fact is that the inner workings of computers, unlike automobiles or telephones or steam train engines, are not comprehensible to the pictorial imagination of the average person; the computer is understandable only in the most abstract and mathematical terms. Not only must the designers be mathematically intelligent, the same is largely true of the repairmen as well; and, as often caricatured, these technical types often lack expressive imagination and human empathy and treat the end user with nerdy contempt. Computerization is an elite field, rigidly meritocratic by its nature, and it requires no immigrant firebox repairmen, no black gandy dancers, and no garage tinkerers with only a high school education.

The new level of technologization in our culture has passed to the point where the difference between what is natural and what is the effect of human technological effort is no longer apparent. It may be that technological improvement is a constant feature of human history, yet it is only now that have we reached the point that technology provides, in effect, all the sensory inputs we experience—meaning that we are now having to make dedicated efforts to search out the natural, including the laws of nature, not to mention nature's creator. The ubiquitous speakers and flat screens surround us like a perceptual blanket, making normal, unaided perception seem bland and washed out. The dramatic intensity of the images on the omnipresent flat screens substitute for the reality of our actual lives, replacing our normal environment with a technologically altered vision of the world. But the flat, colorful images go beyond technological enhancement to actual replacement; they go beyond mere images that provide color and interest to become the reality itself. We look not upon a natural world, or a world in which nature has been enhanced and put to use for the human purposes, but upon a vast technological image that has become our reality.

Previously, I made the point that the typical user of contemporary technology is unaware of the scientific laws that underlay such devices as satellite television and microwave ovens. But this obvious fact (for who outside of electrical engineers and physicists understands in any serious way the science that underlies such technologies?) is only the first step in the transcendence of technological application over pure science. Now serious scholars dispute that there is such a thing as a distinction as between "pure" science and its technological application. Rather, scholars such as Galison and Lyotard argue that the cultural, economic, commercial, and technological contexts of scientific discovery are the essential ground from which such things as Einstein's theory of relativity grow, rather than as the outcome of pure thought and prolonged consideration of recent scientific research.[7] In effect, they argue that scientific theories originate in the practice of technology and are impelled by commercial requirements; this implies that there is no abstract realm of mathematical laws which underlies the physical realm studied by scientists separate from their technological application. With both the examples of Lyotard and Galison present before us, we can see how technology has overcome science both in what the general public expects science to offer (new products and a better, healthier future), and intellectually, how scientific theory is being reduced to scientific practice (not Einstein's intellect but his experience examining patents for electric clocks as the source of the theory of relativity).

From its beginning, modern science has had conflicting rationales to justify it and competing explanations as to what constitutes its essence. Was modern science simply a means for mankind to improve his material condition independent of religious promises about eternal comfort in the hereafter, or was it a new kind of insight into the nature of physical reality independent of philosophy and religious dogma? In essence, was modern science based on careful empirical, experimental observation, or was it an extension of the eternal laws of mathematics and logic shown to impenetrate all the aspects of physical reality? Technology has overtaken these inner doubts and debates in science to settle it that the essence of science lies in practice, not theory, and to imply that that overall science's role is not to understand the physical universe by means of discovering empirically verified scientific laws, but to make practical advances in our ability to manipulate physical reality to our own immediate needs and wants. The loss is that we no longer are likely to understand science as a glorious, intellectual achievement that reveals the secrets and symmetries of nature that can point to a universal designer, but merely as the means to comfortable living.

Notes

1.   Arthur C. Clarke, from *The Quotations Page*, on-line compendium; www. quotationspage.com/quotes/Arthur_C._Clarke.
2.   Lynn Thorndike, *History of Magic and Experimental Science* (1923) is the first history to detail the connection.
3.   Augustine, *The Confessions of St. Augustine,* translated by F. J. Sheed (New York: Sheed and Ward, 1943), 62.
4.   Frank E. Manuel, *A Portrait of Isaac Newton* (Washington, DC: New Republic, 1968). 164.
5.   John Caiazza, "Athens, Jerusalem and the Arrival of Techno-secularism" *Zygon: Journal of Religion and Science* 40, 1 (March 2005), 17–19.
6.   Bryan Appleyard, *Understanding the Present: Science and the Soul of Man* (New York: Doubleday, 1992), 163.
7.   Peter Galison, *Einstein's Clocks, Poincare's Maps* (New York: Norton, 2003), 221–293. See also Jean-Francois Lyotard, *The Postmodern Condition* (Manchester: Manchester University Press, 1984).

Part III

The Dynamic Relationship between Religion and Science

13

Comparative Religion and Scientific Law

False Irenicism

The developments covered in the previous ten chapters make it clear that a rapprochement between religion and science is more likely on the facts than is generally thought. However, in making an approach to the scientific point of view, many religious intellectuals tend to give away too much to science while neglecting the interests of religion. There is a false irenicism that assumes that conversation between scientists and religious believers will eventuate in a successful meeting of the minds with mutual respect on both sides—an assumption frequently contradicted by the statements of the scientific point of view which are often materialistic and offer reductive explanations of religious belief. A noteworthy example occurred when the he Templeton Foundation arranged for a debate between Steven Weinberg, the Nobel Prize winning particle physicist who argues on behalf of a strict and explicit reductionism, and John Polkinghorne, who is both a quantum physicist and a minister in the Church of England. As reported in the *New York Times*, the debate was characterized by Polkinghorne's sanguine belief that religion and science are no longer in serious conflict and Weinberg's blunt declaration that religion is "an insult to human dignity."[1]

Religious advocates have often foregone the dignity of religion by leaping on every neutral or occasionally positive comment about religion from scientists. Thus, if Einstein says the universe should be held in awe, religious advocates (including a very intelligent one such as Fr. Benedict Groeschel) take this as support for religious belief even though Einstein was, at best, an agnostic with pantheist leanings. And scientists will sometimes compare religious experience to the "rush" they experience when a new theory or discovery first occurs to them and while such a comparison may seem supportive of religious belief, it is in fact a condescension. The rush the scientist feels is an emotional response to new knowledge that he or she has attained while religious experience is supposed to be a continuous emotional, spiritual, and intellectual response to intimate contact with the divine. The comparison thus implies that religious experience is as subjective as the scientist's rush and has no connection with objective reality.

The result of too eagerly pursuing reconciliation is that religious belief is redefined in terms amenable to a rationalist and naturalistic point of view which neglects the truth of religious doctrine and lessens its impact and integrity. This can be seen in various ways. First, there is the fact that it is theologians and religious philosophers, not scientists or scientific philosophers, who promote the ideal of ultimate reconciliation. Thus, a prominent scholar of religion and science, Ian Barbour, who writes from a broad intellectual perspective and is sympathetic to religion, posits a four step process in the interaction of science and religion. It begins the with initial conflict between them, then moves to a position in which each exists independent of the other. Next there is a process of dialog between the two and, finally, an integration of science and religion.[2] Barbour's is a historical perspective, starting with the Galileo episode and the separation of the religious and scientific points of view and ending with the projected, but not yet accomplished, future with an integration of religious doctrine and scientific truth. The objection is that Barbour's position appears to give up too much of traditional religious teaching, for writing from a Christian perspective he yields to a process philosophy which tends to "soften" the edges of the doctrines of Christ's divinity and of God's creation of the universe. Process philosophy has often been the recourse of religious intellectuals aspiring to reconciliation, and it is ultimately Barbour's end point as well. However, reliance on process, usually associated with evolution, often leads scholars attempting an irenic reconnection of religion with science to give up such doctrines as the unchanging nature of the revealed God, the divinity of Christ, and the epistemic legitimacy of religion in general.[3]

There is, therefore, a danger in religious attempts to reconcile in that religious believers will cede too much to the intellectual prestige of science. This will lead to acceding on an explanatory level to a naturalistic or materialistic understanding of religious belief—an understanding amenable to scientific explanation but one which at the same time inhibits understanding of the spiritual aspect of religion. However, religion has a legitimate standing of its own in human history as an institution and in its influence on individual human beings that is profound and irreplaceable. Furthermore, religious intellectuals in several traditions have produced comprehensive and plausible explanations of their faiths in relation to secular knowledge including Aquinas, Kierkegaard, and Maimonides. There is a rational case to be made on behalf of religious belief but despite this, like a spouse blindsided by an unexpected demand for divorce, advocates of religion may in the name of a false irenicism give up too much of the rights of religion when, as any good divorce lawyer knows, the best strategy in such a case is to make a set of demands of one's own.

Most of the practicing scientists who accost the issue of religion and science feel no compunction to reconcile, confident that scientific research can proceed without as much as looking over its shoulder at religion except as a phenomenon that is potentially interesting. The actual attitude toward religion of scientists, however, is not often one of outright hostility such as Weinberg's; rather it is one of what might be termed that of the neglectful spouse. While one spouse (religion) seeks reconciliation and desperately wants to reconnect, the neglectful one (science) merely seeks to keep the peace and make a harmonious separation. Thus, Stephen J. Gould in his next to last book, the short elegiac volume *Rocks of Ages* (the title involving a double reference to the "Rock of Ages" in the Christian hymn and the rocks whose ages are determined by geology), argued on behalf of a policy he called "NOMA" or "non-overlapping magisteria."[4] The NOMA policy is intended by Gould to keep religion and science within their appropriate "frames" (a term he borrows from G. K. Chesterton) and not to encroach beyond their limits to try to capture territory belonging to the other. In this way, science would presumably cease from offering reductive explanations of religious belief or characterizing religion as inevitably disappearing because science will expand to explain more and more until religion no longer has a rational function. The *quid pro quo*, it can safely be inferred, is that religious believers will stop attacking science as atheistic and, especially as far as Gould was concerned, stop challenging the teaching of evolution in the law courts or on school committees. (This policy of professed neutrality is the official position of the *American Association for the Advancement of Science.)* Again, this is the attitude of the spouse who is tired of the marriage and who wants only peace to be restored by harmonious separation.

The difficulty with the NOMA policy is that scientists are not likely to abide by it as the recent upsurge in scientific studies of religion by biology and social science indicate. These are mostly reductive and will not prevent religious intellectuals from seeking reconciliation. The human intellect naturally seeks to find unity and coherence in the universe; thus, scientists, religious intellectuals, and serious people in general will seek for the overarching symbol or system that unifies.

One major approach often used by those desirous of effecting a reconciliation between religion and science is to argue that while there is a conflict between *scientism* and religion, there is none between religion and science as it actually exists.[5]

That is, there is a philosophy of scientism that exaggerates the importance of science elevating it beyond its inherent range, which makes of science a virtual idol and subsumes all human knowledge and action to its purview. True science, it is alleged when this distinction is made, does not pretend to be able to answer those questions which are properly asked and hopefully answered in the realms of art, politics, philosophy, history, literature, and religion. It is said, rather, that the materialism, determinism, naturalism, and atheism too often associated with science and made into a philosophy are not promoted by scientists themselves who usually manifest an exemplary modesty. Instead, scientism is promoted by a science's hangers-on and second-rate philosophers.

It is certainly true that if there is to be a reconciliation between religion and science that scientific knowledge cannot be said to imply in a necessary way that the universe is purely material or that the only existent things are atoms and the void. Furthermore, the presumption of this book is that in the mind of God as Holy Spirit, i.e., the manifestation of divine wisdom and of love, there is no conflict because all knowledge is from the Spirit of God within whom there cannot be any contradiction or internal conflict. However, to make the obvious point, we human beings are not divine and, unfortunately, it is not only second rate thinkers who have claimed a scientific vindication for materialism or naturalism but scientists of the rank of Laplace, Weinberg, Monad, Crick and Watson, and E. O. Wilson. And also, among proponents of scientific materialism, are philosophers of the rank of Hobbes, Dennett, and Marx. The point is that there are inherent aspects of science that many serious and highly regarded thinkers have thought does imply a materialist or naturalist philosophy. In the next step such materialism and naturalism necessarily implies the falsity of virtually all religious belief. This leaves religion as, at best, a comforting illusion; a psychological placebo for the ignorant and weak of mind and heart. In order to deal successfully with the arguments of these scientists and philosophers, it will be necessary for religious intellectuals not merely to make the distinction between science and scientism but to deal on some level with science itself in order to discover the inherent incompleteness of the scientific method.

The Final Scientific Law Printed on a T-shirt

The presumed course of science history is that increasing accumulation of accurate data upon which scientific laws are erected eventuates over time in laws of more complexity which encompass more and different types of phenomena. This process continues until, finally, there will be a grand unification in the form of a "final theory," what are known as GUTs (grand unified theories) or TOEs (theories of everything). By discovering the final law, science will have accomplished the greatest feat in the history of human intellectual achievement—for then, science and humanity will know "the mind of God." Physics has recently seemed to be racing toward this final culmination in the form of a theory which would completely describe in consistent mathematical terms the four most basic forces known to physics, i.e., gravity, electromagnetism, the "weak" force among atoms, and the "strong" internal binding force within atoms.[6] This has seemed to be on the doorstep, even though it has not happened so far or as quickly as some scientists had hoped. Yet the culmination of the physicists' project has seemed to be so close that some of them have bragged about it going so far as to say that final formula would be written on a T-shirt.[7]

It is a startling and effective image. The ultimate secret of the universe described not in a divinely inspired religious text or a multivolume history of philosophy but in a concise mathematical formula encompassing all the mysteries dreamed up by the human imagination or encountered in millennia of human achievement and suffering, encapsulated in five or six lines of mathematical symbols easily read off of a casual garment. In the end humanity would be forced to recognize that it is not religion, humanism, tradition, literature, or politics that reveals the last mystery to the human race, but science. It is a challenging image and one worth considering; what would be the consequence if, in fact, the physicists of tomorrow or the day after succeed in providing such a grand and final set of equations?

The answer to this question is bound to be consequential, for if its' challenge cannot be answered satisfactorily then indeed scientism as well as science would be the presumptive "winner" in the religion/science conflict. That is, if that is reason that was to be the deciding criterion. But fideism, or the decision to simply ignore the rational claims of science in favor of an unmovable stance of religious belief, is at best a last refuge since it would give up the field of rational debate to the avatars of scientific method.[8] The fideist strategy strongly implies that the scientific method characterizes all that is rational and that religious faith is the giving up of reason for the security of blind faith. And also that, as some scientific philosophers have claimed, only those afraid to live in the universe as it actually is with its doubts, challenges, and shades of gray will leave science in favor of the comforting rigidity of religious authority. To avoid this result, it would be better to draw the shifting lines in the conflict more to the advantage of religious belief if a compromise or reconciliation is to be attained. So again, what would be the consequence of the appearance of a T-shirt with the final scientific equation?

The answer surprisingly is that shortly after the first appearance of the final formula on a T-shirt, there will appear another T-shirt with a different final formula, then another, and another until it becomes obvious that there will be no final set of equations since science and physics will never settle on a final version. This can be seen in the history of science from the fate of its greatest theory, Newtonian mechanics, and from the mathematical aspect of scientific theorizing, an aspect to which not enough attention has been paid either by philosophers of science or by popular accounts. It is the empirical discoveries of science that are pictorial which grab the popular imagination. Such as fossil tracks in a meteor showing that life exists on Mars, X-rays of brain tissue damage that explain pathological behavior, color photographs of a newly discovered species of parrot in the South American jungles. Or the images provided by scientific theories, such as the evolutionary procession of primate forms from ape to monkey to Neanderthal to man, the bent grid of intersecting lines that defines gravity fields, and the colored balls in repeating patterns of the spiral helix. In addition, finally, there are the iconic faces of the great scientists themselves: the heavy browed and bearded Darwin, the bright and disheveled Einstein, Freud with his glasses and cigar.

But these popular images belie the essence of science—that by close observation and exact experiment it is possible to discover scientific laws that can describe in mathematical form the general pattern of phenomenon. However, general scientific laws are only pictorial as formulas in mathematical form, and it is a fact acknowledged by publishers of popular books about science that every formula in the text loses you one thousand potential readers. Unfortunately, the misunderstanding of science as basically an empirical enterprise is found on the philosophical level as well as philosophers of a materialists and naturalistic bent seem to assume that the example of modern science will automatically make their case. It is noteworthy that philosophy of science, as it has emerged as an academic discipline, in the last fifty years is almost always empiricist or naturalistic. Yet a strong case can be made for science as an "Idealistic" enterprise in the formal philosophic sense because of its reliance on general laws that are mathematically defined. As these laws have developed into greater generality and explanatory power, they have become increasingly abstract, that is, removed to a level nearly independent of the original experimental and observational base.

Over time, it can be seen how newly discovered laws are combined into patterns of increasing generality; i.e., Galileo's law of falling bodies, Kepler's laws of planetary motion, and Descartes law of inertia were incorporated by Newton so as to become derivations from Newton's axiomatically defined three laws of motion. Newton's example is of paramount importance, for as is often stated his mechanical system was overthrown by Einstein's theory of relativity, but it is not often stated how Newton's system was overthrown. A brief look at its end gives the clue. For in the latter part of the nineteenth century Newton's grand mechanical theory had undergone systematic refinement by later scientists, including Laplace, until finally it became possible to understand Newton's system as if it were mathematical rather than empirical in essence. The German scientist Heinrich Hertz, known mainly for his discovery of radio waves, in the late nineteenth century rewrote Newton's system with a refined set of axioms, which made calculation of those laws derivative from the axiom set easier to prove and derive.[9] (Hertz used the physical principle of *least action*, a principle unknown to Newton as a basic axiom.) However, what this last refinement proved really was that Newton's mechanical system had evolved into a form of geometry with a set of axioms and definitions from which laws and propositions could then be derived. However, once evolved into this degree of abstraction then the axioms and definitions could be changed somewhat and a newly derived set of propositions set out. In general, when a scientific theory becomes so fully mathematicized that it can be treated like a formal system, it has escaped its empirical grounding to the extent that it can be reformatted in a variety of formal versions.

Furthermore, within the formal areas of geometry, mathematics, and logic, there has been a virtual explosion of new formal systems including non-Euclidean and four- dimensional geometries, new algebras (including the use of imaginary numbers, binary numbers, and sets), and new logical systems (including propositional logic, truth tables, modal, and more than two value logic). These developments on the formal side have been available to scientists who can now choose the kind of mathematics and logic that best describe their models. The new geometries describe cosmological models including the Big Bang, the new algebras permit mathematically precise descriptions of subatomic phenomena, and the new logics provide the formal basis of both the software and hardware of computers. In fact, there is now a choice of mathematical forms, different axioms, different modes of calculation (including most significantly deterministic forms such as the calculus, and probabilistic forms such as statistical sampling), and chaos theory—which, in turn, gives an insight into physical systems so complex that they escape deterministic mathematical description.

The usual understanding is that the end of the dominance of Newtonian system came about because it could not explain completely new types of phenomena latterly discovered by science including radio waves, radiation, and the perihelion of Mercury. But the continuing mathematical refinement of the Newtonian system also played an essential part in its decline so that what happened in its course of being overcome by relativity theory is that it was subsumed mathematically. In effect, the entire formal system of Newtonian mechanics was made into a subset of a larger, more general theory whose axioms and definitions included the Newtonian principles as special cases. The new set of relativistic axioms and definitions allowed for dynamic relationships between time and space, and matter and energy, which explained certain restricted sets of phenomena that appeared as contradictions to the Newtonian system at extreme speeds or extreme forces. Interestingly, the special cases which Newtonian mechanics are limited to are still the ones that scientists and engineers use in all but a very few applications. Were Newton to visit the Jet Propulsion Laboratory, where the complex paths of satellites are planned in space research, he would learn nothing new on the theoretical level as he was the first scientist to describe the precise force relationships required to project a body around the earth and as the scientists at the JPL use the differential calculus to predict the course of the artificial satellites (the same calculus which he had invented while an undergraduate on break).

With all this in view, summarizing the history and the abstractness of the grandest and most successful theory, what then of the T-shirt prediction? If the T-shirt is based on a deterministic model, another version could be indeterministic, and this is not a mere generalized point since quantum mechanics utilizes a probabilistic form of mathematics and the scientist, in effect, has to make a choice and declare himself on the philosophy of quantum mechanics. Is the action within the atom so difficult to predict that, as at a horse race, we must fall back on probability—or, as Bohr and Heisenberg thought, is the probabilism inherent within the atom itself? Alternatively, is it possible to impose a deterministic pattern on quantum phenomena as Einstein and the physicist-philosopher Bohm thought? The point is that a choice in the matter of quantum physics produces materially different sets of final equations. And even when there is no difference in theoretical commitment, there can be a variety of choices of mathematical forms used to describe the same phenomena. As Feynman points out, there are three kinds of equations that can be used to describe the attractive force between to physical bodies.[10] Thus, the first T-shirt imprinted with the final theory is guaranteed not to be the last.

God and Nature

A conflict between science and religion cannot arise if there is no essential difference between God and the physical universe. The reason is transparent, for if God is coexistent with the universe in a pantheist fashion, then a conflict between religion and science is metaphysically impossible because it would be a conflict within a uniform divine-material reality. Mixtures of the divine and the earthly that do not separate them are, in a way, appealing and natural to the human imagi-nation—a fact which explains the polytheism of ancient culture as well as the multiplicity of manifestations of the divine in Hindu worship and perhaps Catholic veneration of the saints. The pantheism of Spinoza and the monism of William James are reversions to this appeal and, again, transparently eliminate the source of the conflict between religion and science. But Hinduism and philosophical pantheism are not the major religious influences in Western culture and so we must examine the effect of the revelation of the duality of creator and creation. While the conflict between religion and science becomes apparent in Western civi-lization after the beginnings of Christianity, even before then it appears in a nascent manner in ancient Greek and Roman culture. This was a time when the beliefs of the common people in their many gods, often more human in appearance and behavior than godlike, were criticized by philosophers of the time. Thus Xenophanes, not often enough cited as one of the major pre-Socratic Greek philosophers, said pointedly that while the Greek gods were white skinned, African gods were black, and mankind generally imputed human characteristics to the nature of the gods. But rather than concluding that there were no gods, Xenophanes, like Socrates, Plato, and Aristotle, stated that God's nature was not like that of mortal men but that he was immortal and he had no sense organs; Xenophanes asserted that instead of many gods there was in fact only one God.[11]

The importation of the Hebraic concept of divinity into ancient European culture took the form of the Christian message, which could be viewed as an acceleration of the development of ancient thought from polytheism to monotheism. It continued as a refinement of the idea of the

nature of God that gave a personal character to the abstract characterizations of the philosophers. There was acceleration also in the portrayal of
the relationship between divinity and the universe, namely, the biblical
concept of creation that had the effect of inserting a sharp separation
between God as creator and the universe as creation; the creation subject
to the Creator.

The origins of the conflict between religion and science in the modern
and the postmodern situation can be seen in the sharpness of the separation between the physical universe and the divine, as creation and creator
respectively. An aspect of the separation is the difference in how we
come to know the physical universe and the spiritual aspects of reality.
In modern terms, since the Enlightenment the classic scientific method
of close observation, experiment, repetition, measurement, hypothesis,
and testing became the means by which we know the physical universe.
This gave us a firm and eventually settled knowledge of the causes and
relationships of physical phenomena. Scientific method implied a guarantee that the universe was in the grasp of the human intellect and firmly
in human control. Thus, this meant that scientific rationalism was not
merely hubris, but a reflection of the success of the scientific method
both in terms of the increasing complexity of scientific theories and in
the usefulness of scientific knowledge in its applications to industry,
commerce, transportation, communication, and health.

By contrast, revealed religion is based not on rational methodology but
on faith. Indeed, in the words of St. Paul citing the actions of Abraham
as justified not by his works but by his faith, and in the actions of Jesus
in the Gospels, faith was the means by which to know and understand (to
the degree that it was possible) the nature of God and of spiritual reality.
Faith, according to St. Paul, is knowledge of things unseen and therefore
beyond immediate proof. Now again, this did not mean that the revelation was beyond rational discussion, but rather indications of the truth
of the message were given in indefinite terms. They were manifested in
the occasional miracle and healing, the history of Israel and the Church,
and the political situation of oppression and domination. Additionally
they were apparent in terms of the personal experience of believers (e.g.,
feelings of despair or unworthiness) and in protest of the gross immorality
of the popular culture; these themes are apparent in the writings of St.
Paul. They continue to be the basis of the appeal of revealed religion in
the twenty-first century. From the epistemological to the metaphysical,
we proceed to remark, finally, that religion understood by means of the
Hebrew revelation and its progeny in Christianity and Islam separates

the creator God from nature as His creation (the personal nature of God never in doubt). Moreover, the conflict between religion and science has its ultimate cause in this separation.

I have been speaking of the conflict between religion and science thus far as if "religion" had a restricted meaning, in effect, that religion means worship of a spiritual entity that is distinct from the physical universe. However, postmodernists and others would protest that such a definition is too narrow, that there are many religions to be found in the cultures of the whole planet and within human history, and that their belief systems cannot be so readily categorized. The yin and yang of traditional Asian culture provides a template for the resolution of the conflict between religion and science by providing an image that combines opposition with balance. The insight from Indian religion that the way to enlightenment transcends all conflict both within and without the human person also provides a means of encompassing the science/religion conflict. It might, therefore, seem a convincing scenario that by importing these insights we can do away with the conflictual notion of war or opposition between religion and science.

In response, it can be argued that the revealed or Abrahamic religions yield a precision that makes intellectual progress more likely in both science and religion. Modern science is largely, even if not solely, the creation of Western culture and sprang from the notion that physical nature worked on its own independently of magical interventions—a notion which developed over time into the network of separate scientific laws by which to understand and control the physical universe. The Abrahamic, or revealed religions, are preeminently theistic and give a definition to the concept of the divine separate from the human and thereby less likely to be infected by human imagination. Hence, the conflict between religion and science arising in the West remains a potent and defining issue, which the importation of the concepts of Eastern religions is not likely to affect. On the other hand, concepts such as holism and multiple and overlapping forms of causality that arise within the more speculative areas of contemporary science may provide a means of overcoming the conflict.[12]

In the meantime, it should be pointed out that the reductive motive of science has not evaporated despite fervent hopes otherwise, and in the hands of Wilson, Dennett, or Weinberg, religious worship generally is reduced to the resultant of various physical causes naturalistically understood. Apparently, there is nothing so exclusively human in the face of the reductive urge that it cannot be shown to have a genetic cause or a subatomic basis—such that the teachings of Buddha, the Vedas, Moses,

Jesus, and Mohammed are all dissolved into fruitless human imaginings with no basis in reality.

Revealed Religion and Scientific Law

As is apparent from their history, it is the revealed or Abrahamic faiths that experience a conflict between their doctrines and the putative discoveries of modern science. It is worth exploring further the attitude of the Abrahamic faiths in chronological and religiously phylogenetic order: Judaism, Christianity, and Islam. Since there are differences within each of the separate strands of the revealed tradition as they approach the physical universe, it is worth offering a brief impressionistic summary of each. In effect, how do each of the Abrahamic faiths overcome the tension between faith and reason since they all offer specific features which seem to represent unreason in supreme conflict with rationality? Moreover, how does that affect how each of them understands scientific law?

Judaism; The Lord as Moral and Physical Law Giver.   The seeming irrationality of the ancient Israelite tradition is represented by the image of a bloodthirsty Jehovah, for which examples are readily at hand. Critics often cite his commands to destroy or kill whole villages. Indeed, Saul, the first king of Israel, lost the confidence of Jehovah when he refused his command to kill every living thing in the kingdom of Amalek including children, women, and cattle. (1 Samuel, ch. 15) The prophet, Elijah, in competition with the priests of Baal, wins by making a miracle that the competition cannot match—the result of which is the slaughter of 450 pagan priests. (1 Kings, ch. 16) These accounts certainly are repugnant in the postmodern age when violence is seen as the result of failure to communicate, and toleration is understood to be the most basic social virtue—nor is this repugnance only the reaction of a contemporary post-modern sensibility. While in the postmodern world recourse to violence is seen as the negation of the social order, in former times it was understood that the use and the readiness to use violence on criminals and enemies of the state was an essential condition of sustaining social life. Violence, or the threat of it, is still essential to the social order but that fact tends to be covered over or disguised. Thus, we are reluctant to use the death penalty even in the most heinous cases of cruel multiple murders, but we accede readily to a very high rate of imprisonment of young men as the acceptable cost of keeping our streets safe.

The bloodthirstiness of Jehovah in the Hebrew Bible has a consistent logic, however repulsive the violence. This is because, in the cases recounted above and in every other instance, the command of Jehovah to destroy Israel's enemies was intended for two evident purposes: first, to provide a secure land for the nation of Israel and second, to maintain the worship of Israel purely dedicated to the Lord God. In an ancient setting where polytheism was the cultural norm, the command given by Moses that Israel's God was one (1500 B.C.) was not readily or consistently carried out. It is not until the time of the prophet Ezekiel (600 B.C.) that Israel appears to have been unified around the worship of one God. Saul was commanded to destroy an entire village and all inhabitants to punish a kingdom whose attacks had almost destroyed Israel. Elijah's mission was to keep their worship of false gods from contaminating Israel's worship of the one true God (a mission that had to be given to succeeding prophets as well), which required a violent face-off since either Elijah or the pagan priests would be killed. Thus, the violence and the commands to kill may be objectionable as a means of policy, but they nonetheless reflect a consistent rationale.

But why was the religion of the Israelites more important than those of its surrounding neighbors—Assyrians, Greeks or Philistines? Is monotheism worth the price, we may ask. The answer is that the one God who is revealed in the Hebrew Bible appears as lawgiver to Moses and the Israelite people in the Ten Commandments and the hundreds of other laws, prescriptions, and prohibitions in the early books of the Bible (Numbers and Deuteronomy). But the Lord also asserts his claim to be the lawgiver to the universe in the physical sense as well as in the legal or religious sense first as Creator (Genesis, ch. 1) and then consistently through miraculous interventions—as when Joshua was able to command the sun to stand still during the battle at Jericho. Later theologians would perceive in the announcement to Moses, "I am who am," that God was the ground of universal being, but, in the Hebraic context of that time, it simply meant that Jehovah was the giver of all laws in every aspect of human life and physical reality. It was in this sense that monotheism gave, and gives, a radical coherence to the believer's understanding of the universe and life as it is experienced. In those ancient biblical rules that insist on punishment for transgression and the texts which praise the Lord for his providential concern lie also the sense that human life has a connection to the divine—a sense that provides a meaning for human existence as reflected in the round of daily observances of the observant Jew.

Christianity; Pursuing the Incarnation.   The great challenge to reason from the Christian religion might seem to be in the claim that miracles occur, such as the cures at Lourdes but especially Jesus' rising from the dead. However, miracle accounts are found in virtually in all religions, and the Hebrew Bible recounts two raisings from the dead performed by the prophets Elijah and Elisha (1 Kings, ch 3, 2; Kings, ch. 4). There are also examples of return from death in accounts of gods such as Apollo, who dies in the winter and arises again in the spring. Nor is the main challenge caused by the Christian claim that Christ was divine since many people have claimed to be divine or to share an intimate connection with the divine, such as Mother Annie Lee, Joseph Smith, and Mohammed. What is uniquely challenging to scientific reason from Christian religion, in fact, is the doctrine of the Trinity.

The doctrine of the Trinity is blatant in its challenge to reason, for it asserts that there are, to use the formula from the Council of Chalcedon, three persons in one God. In mathematical terms, *three equals one,* thus denying frontally what Aristotle considered the most essential condition of reason, the principle of identity. So blatant is the challenge to reason that many Christian believers have attempted to do away with the doctrine, including Isaac Newton, and in this manner attempted to make Christianity rational and preserve the most appealing parts of Christian belief. There have been many such attempts including those by the Arians in the fifth century and the Unitarians in the nineteenth. Also, what can be termed "post-Trinitarian" religions have arisen, including Islam and Jehovah's Witnesses, which claim to be rescuing the nature of God from the contradictoriness inherent in the Trinitarian doctrine that seem akin to polytheism. However, there is a further logical aspect of the doctrine of the Trinity besides its self- contradictoriness, namely, that it is a consistent way to maintain the divinity of Jesus Christ who is the defining object of Christian worship.

The connection of the Trinity with the doctrine of the Incarnation arises from the fact that Jesus is a human person, yet divine; he is one with the Father, yet by the observation of his apostles and others, a separate (human) being. By eliminating the Trinity, post-Trinitarians at the same time by logical necessity must say that Jesus is less than divine. This is indicated by their various categorizations of Jesus as a prophet (Islam), an angel (Jehova's Witnesses), a holy man, a great ethical teacher (Unitarianism), a mythological figure, a lesser divinity than the Father (Arianism), or an appearance of the Father on earth but without a true body. Eliminating the Trinity negates the identity of Christ as one with the Father. (St. John, ch. 14, 8–16)

But is preserving the divinity of Christ so that Christians may feel secure in their worship worth the price, we may ask. The answer seems to be that preserving the divinity of Christ is necessary for maintaining the doctrine of the Incarnation; a consequence of which is that the Christian is able to share in the aspects of divinity manifested on earth. The Christian believer does this by acting in the manner prescribed in the Gospels, manifesting divine love in an outward manner to others in the Christian community but also by attempting to discover the various signs and evidences of divinity in the natural universe. The discernment of divine patterns and signs provides evidence for the faithful believer of divine presence in the physical world, an aspect of Christian belief that impacts the understanding of physical laws. Physical laws, for the Christians who are scientists, are manifestations of divine presence not because the laws are miraculous—for by definition miracles occur when physical laws are broken and the chains of scientific causality interrupted. Rather, the patterns themselves are evidence of design and purpose since they constitute the lineaments by which the creator made the universe—the mind of God as it were. Christians who are scientists are explicit on this account: for them, scientific discovery becomes a derivative form of Christian worship and their scientific research, a vocation. This includes Galileo, Kepler, Newton, Mendel, Faraday, and Maxwell.

The status of physical law in the Incarnational view is that they have a reality of their own—the network of scientifically discovered laws a testimony to the creative aspect of God. Incarnation means not the subsumption of physical reality into the spiritual realm as if it were a reverse form of reduction, but a respect for the reality of such laws in themselves. The nature of physical law, after all, is that while they describe and define the behavior and nature of physical entities, the laws themselves are not physical but abstract so that their metaphysical constitution does not pose a threat to the concept of spiritual reality (see chapter 13, section 2 above). This respect for the reality of physical law is reflected in the respect accorded in Christian doctrine by Jesus and St. Paul toward the laws of the civic order as well. Jesus famously says, "Render unto Caesar what is Caesar's and unto God what is God's," and St. Paul instructs one of his communities not to cause aggravation to the kings of this world since because of their divinely ordained responsibility they have to answer to God in the next (Matthew, ch. 22, 21; I Timothy, ch. 2, 2). For the Christian believer, the world is real in itself and can be taken on its own terms on a qualified basis as long as due deference is given to the presence of the divine. Thus, this enforces a kind of humility as if the world order is real but under the lordship of the almighty because it was created and is sustained by God.

Islam; Physical Law versus the Divine Will.   Islam is a post-Trinitarian religion and its long history and huge number of believers worldwide testifies to its importance. It is a revealed religion as its object of worship is the God of the Hebrew Bible. Islamic scholars invented the term "Abrahamic faiths" to indicate Islam's relationship to the Jewish and Christian religions. The irrational aspect of Islam is apparent today in the political realm in the attacks on 9/11, on xenophobic insistence on maintaining smaller Islamic islands of faith and practice within Western states, and in general in the outright ferocity of Islamic fundamentalism in defending and expanding itself in such a way as to make compromise impossible. Islamic fundamentalists see no reason why believers cannot live on the welfare state benefits of Western nations while at the same time working to undermine them, or why they should not vigorously claim the rights of freedom of religion and expression in Western states while denying them to non-Islamic believers who live in Islamic states.

The irrationality of political Islam is only apparent, for its attitudes and policies follow directly from its theology. It is not enough to say, as is often said in the postmodern mode, that fundamentalist Islamic policies make sense to them and they have as much right to them as believers in any other faith have to their own seemingly irrational doctrines. It is, rather, that the ferocity in defense and expansion is necessary consequence of their devotion to Allah, the Islamic version of God. The Islamic concept of God, however, is different from the Hebraic concept from which it is descended. While the Jehovah of the Israelites was severe and demanding, his followers were encouraged to have faith that all the events of Israel's history had a cause and would lead to the fulfillment of His purposes and to the preservation of Israel. Much was left to the Lord working His will through history, such that the alien empires that were Israel's enemies were working to God's purposes unknown to the emperors and kings who seemingly worked their will on Israel—but, of course, to understand this required great faith. This is not so with Allah, for while he demands obedience, it is as if His will can only be fulfilled by the most active and devoted behavior of his worshippers. There is no relaxation in this view as long as Islam is contradicted in any way. The ferocity with which fundamentalist Islam meets the non-Islamic world is not indicative of irrationality, for it based on its doctrine of monotheism insisted at great price and effort by its believers as a means of cleansing the world from idolatry.

As indicated above, Islam's theology is an attempt at a rationalization of Christian doctrine, particularly the Trinity. By denying the Trinitarian nature of God, indeed, by characterizing it as a form of polytheism, Islam necessarily denies the divinity of Christ, relegating him to the status of a prophet, albeit a great prophet second only to Mohammed. The denial of the Trinity has a result of making a place for Islam's prophet and founder, a world historical figure who combined the talents of a great religious prophet with those of a military empire builder; Zoroaster meets Julius Caesar, Joseph Smith meets Napoleon. Governments do not have credence on their own, and no prophet of Islam would ever give the kind of advice that St. Paul gave the early Christians to obey their kings, in effect, independently of whether or not the kings were Christian believers. Thus, governance structures in societies dominated by Islamic believers are never settled or secure since at any moment an inspired Muslim cleric can declaim against it, attacking its legitimacy on the authority of Islam's holy book, *The Koran*. So much is apparent in today's news accounts from the Middle East; what is less obvious is that this same attitude applies to scientific law.

It is true that Islamic society produces scientists and technicians as good as those from Western nations, that Islam made unique and important contributions to early modern science, and that the Koran is filled with testimonies to the beauties of nature. However, the philosophy of science implied by Koranic doctrine implies that scientific laws are not to be regarded as patterns that inhere in the structure of the physical universe. Thus, the network of laws that explain the behavior of the physical universe envisioned by scientific reason is deemed not to exist and, in fact, condemned as a form of idolatry (Koran, ch. 37, 88–96). The patterns discerned by human reason, including not just exotic scientific discoveries such as the Carbon cycle but more homely aspects such as the change of seasons, are explained not by a subsistent underlying cause such as a general law of nature but at each instance, as caused directly by the will of Allah (Koran, ch. 80, 25–29). There is an atomicity implied in this view, reflected in Islamic philosophy in the influential writings of al-Ghazali, of time as a collection of instants; for example, each instant is upheld and caused by the divine will with no inherent connection between the instants of time. The will of Allah, thus, forecloses the possibility of the inherent causal capacity of scientific law and literally condemns determinism or any view of scientific laws such as the "iron laws of necessity" because such doctrines remove the glory from Allah by denigrating the activity and force of his will. In Islam the separation between God and his creation is complete, the subservience of the natural world in its political, moral, and scientific senses created and overseen in every complexity and detail by the judgmental eye and absolute power of the divine being.

Notes

1. Carey Goldberg, "Crossing Flaming Swords over God and Physics," *New York Times* (April 20, 1999), D-5. See also Vincent Kiernan "Can Science and Theology Find Common Ground?" *The Chronicle of Higher Education* (April 30, 1999), 17, 18.
2. Barbour, *Religion and Science*, 77–105.
3. Barbour, Ibid., 209, 214. Cited by John Polkinghorne, *Belief in God in an Age of Science* (New Haven, CT: Yale University Press, 1998), 40.
4. Gould, *Rocks*, 125–170.
5. John F. Haught, "Science and Scientism: The Importance of a Distinction," *Zygon* 40, 2 (June 2005), 363–368.
6. Weinberg, *Dreams*, 221–241.
7. The image is attributed to physicist Leon Lederman; see Dan Falk, *Universe on a T-shirt* (New York: Arcade, 2005).
8. Bryan Appleyard seems to approach a kind of secular fideism in *Understanding the Present*, 231–235. See my review: "Science Versus Theology," *Modern Age* 37, 3 (Spring 1995), 257–260.
9. Heinrich Hertz, *The Principles of Mechanics Presented in a New Form* (New York: Dover, 1956).
10. Richard Feynman, *The Character of Physical Law* (New York: Modern Library, 1994), 49, 50.
11. Xenophanes in Philip Wheelright, *The Presocratics* (Indianapolis, IN: Bobbs-Merrill, 1982), 32.
12. Polkinghorne, *Belief in God*, 59–67.

14

The Dynamic Relationship of Religion and Science in the Postmodern World

The reintegration of science and religion under the rubric of a grand philosophic synthesis is not on the intellectual horizon in the form of a master narrative. There have been several proposals made by religious intellectuals, including some who are scientists, for an overriding or overarching category of explanation that would subsume both religion and science in such a way as to leave the credibility of religion, particularly the revealed religions, intact. Theologian Wolfhart Pannenberg, for example, uses the concept of a *field* (borrowed from physics) as a means of integrating Christian theology with scientific thought. Teilhard de Chardin famously posited an "omega point" at the end of evolutionary development that coincided with the Second Coming of Christ.[1] At the present time, however, the best that can be hoped for is not a complete or even nearly complete resolution of the issues that lie between religion and science but (as has been suggested in chapters 3 though 12) a realization that the disagreements are not so sharply divisive as many have previously thought.

We are left with the realization that while there exists the possibility of an overarching theory, none are likely at this point to appear intellectually compelling, and within the context of postmodern culture the fact that religion/science intellectual combinations are possible is probably irrelevant. This dire situation arises from the fact that postmodern thought rejects master narratives on principle and would not be interested in the possibility of integrating two of them. And to repeat the point made in the first chapter, postmodernism's philosophy of epistemological relativism undercuts the truth claims of both religion and science.

The relativism of postmodern thought extends from its ironic epistemology to its make-your-own metaphysics, to its promiscuous aesthetic, and it extends to its ethical philosophy that can be fairly characterized as antinomian. The postmodern world is vacant of publicly acknowledged values and moral rules, except at the most extreme end. Thus, to generate a sense of moral outrage, it is necessary in the popular culture to look to the worst possible examples of human cruelty and depravity: child molestation, serial murders, and the Nazi death camps. Whereas child neglect, common murders, and arson fires run past our attention spans in a constant stream on the flat screens that surround us and no longer excite our outrage. Actions thought worthy of moral criticism just twenty years ago are now acceptable in postmodern culture, including insulting teachers and parents and, as pointed out later, pornography. In a somewhat Marxist sense, there is a material basis underlying the relativism of the postmodern world, particularly its moral relativism, which is its most repugnant and frightening feature. The material basis of postmodern ethics and the relaxation of moral rules is the new level of technology that has seriously altered our common lifeworld when combined with the utilization of the new technologies by commercial enterprises to sell products to the middle class consumer.

Technological Ethics

The impenetration of the contemporary lifeworld by technology has evolved a certain style of ethics which, while not so far written into an explicit theory, has detectable characteristics. Technological ethics is a kind of ethical egoism, more passive than the egoism of Ayn Rand's philosophy of objectivism, but secure nonetheless in the idea that the human individual has no inherent obligation to anyone else or to society as large. Even so, there is an inherent expectation that the human individual needs to depend on an array of supports and maintenances, and these should be made available as a matter of right. This expectation is not perverse, even though it means that the individual takes, as it were, from society without having the responsibility of returning anything back. It is based on the efficacy of contemporary technology, which has increased agricultural and industrial productivity to an astounding degree, and on technology's perceived ability to fulfill every human need, lifestyle want, and personal desire. Thus, the individual in technological ethics is a kind of drone and not the active entrepreneurial and defiant freebooter of Ayn Rand's fantasies. To make the analogy more precise, the techno-secular individual is not only a drone but a drone who expects to be treated like a queen, pampered but without the obligation of producing offspring.

Technological ethics takes a wider view of human good than just what is pleasurable and, in classical terms, is eudaimonistic rather than hedonistic, but it is nonetheless devoted to the notion that the good life means feeling good and looking well. Self-empowerment and self-fulfillment are its keystones to a good life, and what elements constitute a good life is closely defined by technological options. That is to say, not children or family life necessarily constitutes the good life, although these are treated as possible options, nor is a life filled with pleasures and excess devotion to pleasures such as smoking tobacco products and drinking alcoholic beverages considered the good life. Career and a sense of accomplishment are an important part of what constitute a good life, and the pursuit of pleasure can get in the way of these things. Ultimately, technological ethics is limited by its material horizon even as that is a technological rather than a natural horizon.[2]

There is one pleasure that techno-ethics does not stint on, namely, sexual activity which again relies on a technological interface. Sexual orgasm is the most intense single sensory experience a person can know and once removed from its social packaging of modesty, awe, morality, and family life, there is no ethical principle which prevents the pursuit of experience of sexual behavior at any time or of any variety. The polymorphous perversity which Freud maintained is at the heart of infant sexuality now becomes realized in adult behavior. The role of technology is paramount; birth control and abortion separate normal sexual activity from procreation, while antibiotics inhibit the spread of venereal disease, and a variety of sexual devices are made available to aid in what was formerly understood to be perverse forms of sexual activity.

Technological ethics is *instrumentalist* in that what is acceptable as ethical behavior is defined by what is technically and scientifically possible, without any admixture of nonmaterial ethical principles. Indeed, the attempt to interpose moral or religious objections to the use of new technologies is seen as impertinent and obstructionist, as in the case of infant stem cell research.[3] Here, President Bush's opposition can only be seen as motivated by an attempt to satisfy the demands of his political base whose objection derives from strongly held religious beliefs, beliefs which because they do not depend on scientific evidence, are presumed to have no inherent rational justification. The infant stem cell controversy shows the difficulties in posing a moral objection to a technological advance where the instrumental effects are likely to be beneficial, and where the practical, material payoff is thought to be so large that any moral objection seems foolish.

Technological ethics has a link to one particular traditional ethical theory, namely, utilitarianism. The implied argument is that the eudaimonian good of a comparatively large number of individuals (as opposed to their pleasure in the traditional version of utilitarianism) overcomes not so much the denial of good to the few, but the idea of any inherent rule or guide not based on utilitarian terms. In the stem cell debate, the opposition of ethicists such as Leon Kass is seen as obstructionist rather than principled because of the promised results of experimentation on the cell. That human DNA abstracted from a zygote is used simply as a medical substance does not have the effect of freezing the technologically defined conscience because, in technological ethics, the ends justify the means, and the means have been greatly expanded in the new range of reproductive technologies now available.

The most significant effect of new levels of technology on ethical decision making is to have relieved the negative consequences of immoral actions, for postmodern ethics depends upon on technology to obviate the bad or unwanted consequences of our actions. In the past, the great barrier to recreational sex was the possibility of pregnancy, but now with the easy availability of birth control and abortion, the consequences can be managed and the possibility of pregnancy not seen as a threat. Using the example of sex to make the point may seem predictable, so a better example is that of vandalism, considered as the casual breakage or damaging of another person's property. Here, the use of spray paint cans to mark one's territory, the heaving of a brick though a plate glass window, or "keying" an automobile are typical examples; what has also become typical is that such acts are treated as minor crimes. The causes as well as the social results of such laxity have been argued by criminologists, but one of the contributing reasons, surely, is the increased efficiency of contemporary levels of technology.

Consider when a store manager discovers in the morning that a plate glass window has been shattered by a vandal overnight, his first response may well be to call the insurance company. Once informed, the company will contact a glass replacement service it has under contract to take out the shards of the old window, clean up the mess, and replace the window with a new sheet of plate glass. Only afterwards are the police called to report the incident since they are not at all likely to catch the vandal and can do nothing to help defray the cost of the window replacement. There is not likely to be much emphasis on policing window breakage since it is a fairly minor crime, and replacing the window is something that technology supplies almost painlessly. Plate glass is produced efficiently by high-tech machinery and is currently produced in such quantity that it is readily available. Property insurance is widely available because of the constantly improved efficiencies in the way offices, commerce, financing, and communication are supported by technology. The store manager does not have to write out a check to the window company because his account with the insurer defrays all or most of the cost of the window replacement, and any cost not covered by insurance is deducted from his store's account electronically. The store manager will receive a detailed financial report about the incident on hard copy by mail or more likely electronically by e-mail. His report of the incident may require paper forms and a police report for verification, but even that may take place after the fact, only after the plate glass has been replaced—often within a day.

Consider, as an alternative, what would happen if plate glass were so expensive that it would cost the store a good deal of money and would not be available for many weeks. Then, the manager's first call would be to the police, and he would expect them to take serious and sustained action to discover who the vandals were, have them arrested and, if possible, force them to make restitution. The episode would be taken much more seriously, not as youthful high jinks or as an urban art form, but as a serious crime because of cost. As for the vandals, the social effect would be that they would be seen as destroyers of another's property and treated as criminals. More importantly for the vandals themselves, this might force them to consider such actions more seriously beforehand, force them to think about the severe consequences if caught, and force them to develop a conscience about such actions. This mind experiment indicates the effect of technology on common morality—that in removing the effect of the consequences of immoral actions the inherent sense of the evil associated with such actions is thereby relieved or eliminated. With technicalization, a whole range of hitherto evil acts are reevaluated according their no longer bad consequences and are now no longer considered evil in themselves.[4]

The traditional understanding of the difference between good and evil as categories that apply to human behavior is simply vaporized in the technological view. In technological ethics, ultimate good does not derive from nature or revelation, but from mankind's ability to change nature to its ends and desires, and, in this way, human nature acquires a new characteristic which makes it distinct from traditional ideas of human nature. Technological mankind is more than a toolmaker, or *homo faber*, and has now assumed in some part the quality of a *magus*, of a magician who can bend matter by strange and powerful means to whatever ends desired. The good rests in the magical ability to change nature by technological means, to manipulate and live in a purely self-created environment.

Pornography as an Example

How well does postmodernism, considered as a social fact, deal with moral decline? After all, in the postmodern world, moral decline is often disguised as social change, and social change is seen as a means to increasing diversity while respect for different cultures prevents the kind of outrage or judgmental attitude required to make prescriptive moral judgments. Yet, at the beginning of such slides into new abysses of cultural decline, there is always a sense of (if not outrage) unease; it is understood that inevitably the new movement will proceed energetically until it becomes an accepted part of the cultural landscape. This happens in the way of the frog in the soup; the soup starts at room temperature but is increased slowly to the point where the frog is boiled alive. What begins, for example, as a concern for women who become pregnant as victims of rape or incest, in time becomes a fully articulated right to an abortion and given constitutional protection. Up to now, the energumens of social change have been successful, but there can come a point where the destructive presence of such movements in contemporary culture is no longer avoidable and a counteraction arises. When the negative effects of the behavior become apparent, the former sense of moral reality begins to reassert itself. Thus, one writer claims that the "pro-choice" side has lost the abortion debate because while fully legal, the general culture now assumes that abortion itself in inherently immoral.[5] While several examples of the restoration of moral sensibility in the face of moral decline may be cited, the most obvious one these days is pornography.

Regarding pornography, the first thing to realize is just how far we have come. The nudes featured in *Playboy Magazine* forty years ago now seem like portraits of calm feminine beauty, almost desexualized so that the experience of looking at them as we turn the slick pages is like looking at the nudes in classical paintings. But sexual titillation is the true emphasis in pornography, not the portrayal of female beauty or, as is often claimed, to encourage the contemporary culture to take a more relaxed and healthy view of sexuality. And as we get used to images to the point that they no longer have the same effect, an intensification of the image is required to obtain the same level of physiological or emotional excitement and so the images get more intense over time. As long as the vehicle for pornography was the printed page, its distribution was limited and could be controlled to a large degree; but now technology has enabled the spread of pornography to every television, computer, and personal communicator screen making limitation and control nearly impossible. Pornography is a constant component of the recent explosion in electronic technology, and the effect has been to sexualize the culture in its essence so that females are posed not so much to expose their flesh as to incite sexual desire and that not of the most desirable kind.

Hard-core pornography has become an accepted part of the entertainment industry and a huge business. Sexual scenes and innuendo are a constant part of popular culture, including children's cartoons and girl's fashions. Mothers raised in the 1970's, when sexuality was escaping fully from Victorian prohibitions, now complain bitterly that their preteen daughters want to dress like hookers. Also, some high schools have had to enact explicit policies prohibiting the wearing of T-shirts that brag that the wearer is a "pimp." The general culture of the postmodern world has become sexualized to the point where, finally, the ubiquity and destructiveness of pornography is being criticized not only by religious moralists but also by secular critics and, more significantly, by social science.

Resistance to pornography and the sexualization of the general culture was at first the province of religious moralists; evangelical Protestant pastors or Catholic bishops would decry its presence in a popular film or advertising campaign and organize a boycott. Usually these efforts would fail, characterized as a rigid and unhealthy reaction, a threat to First Amendment freedom, and as a naked attempt at the imposition of a theocratic form of government. The religious critics were dismissed out of hand even as the sexualization proceeded; at one point the late Steve Allen, well known as a pioneer television entertainer but also as a serious thinker of liberal persuasion, condemned television in the 1990's as a cultural "sewer."[6] There were congressional investigations into pornography, not a subject that Congress would willingly concern itself with except for the pressure of constituents. However, there came a turning point when resistance to pornography and condemnation of sexualized culture became legitimate and no longer the province of religion and scattered episodes of public indignation—that was the sex scandals of the Clinton administration. The impeachment of President Clinton was, as his supporters claimed, about sex and not that he lied under oath. However, that only meant that moral evaluation of errant sexual behavior was now permissible in postmodern culture. Criticism of the president's sexual behavior could be characterized as cynical partisanship, but his acts were so regrettably offensive that many of his supporters had to condemn them as well. The next turn in resistance to the sexualization of postmodern culture has recently arrived with the notice of the mainstream press and, above all, with the research of social scientists.

In late January of 2006, an article appeared in The *Boston Globe* entitled "The pornification of America" and was offensively illustrated by a portrayal of the Statue of Liberty as a pole dancer.[7] It recounted how "From music to fashion to celebrity culture, mainstream entertainment reflects an X-rated attitude like never before" and gave several examples. The article also cited, among other critics, Professor Barbara Dafoe Whitehead, "We have an aging society and an adolescent culture." Several days later, indicating how the topic of sexualization has now attracted the attention of serious researchers and social scientists, a long piece appeared in the *New York Times* which cited not only social critics but the research of social scientists about the effect of sexual imagery on teen behavior.[8] Among the research cited was a July 2005 study in the journal *Pediatrics* which had been requested by Congress and supported by the federal Center for Disease Control and Prevention. It attempted a correlation of the sexualization of the media including television, movies, and the internet, with the disturbing negative effects of early sexual activity among adolescents that included pregnancy, sexually transmitted disease, depression, and suicide. Television programming is now being systematically examined to correlate the degree of sexual imagery and innuendo with the attitudes and behaviors of teens in studies usually done by means of surveys. One study stated, "the shows most watched by adolescents in 2001-2002 had 'unusually high' amounts of sexual content compared with TV as a whole" and that "characters involved in sexual behavior in TV programs rarely experience any negative consequences."[9]

Prior to the appearance of these studies, the American Psychological Association had set up a "Task Force on the Sexualization of Girls" in 2004. In early 2007, the APA released an executive summary of its ongoing study based on numerous research studies and journal articles. It found ample evidence of "children [especially girls] imbued with adult sexuality," in television, music, movies, magazines, the internet, and advertising. Summing up the consequences of the sexualization of girls, the study stated, "Psychology offers several theories to explain how the sexualization of girls and women could influence girls' well-being. Ample evidence testing these theories indicates that sexualization has negative effects in a variety of domains, including cognitive functioning, physical and mental health, sexuality, and attitudes and beliefs." Done in a scholarly manner, prescinding from overstatement perhaps, the study seems reluctant to make the more general point that the popular culture is imbued with vulgar sexuality to the detriment of all but particularly of those most vulnerable.[10]

The negative effects of pornography are now a matter of scientific record and, having been noted by science, no longer debatable. The negative effects of pornography and sexualization were apparent long before scientific notice was taken of them, however. The first and most prominent critics were Christian moralists, but they were joined by feminist critics such as Cynthia McKinnon and Andrea Dworkin who condemned pornography for its "objectification" of women as sex objects. The combination of religious and feminist condemnation of pornography was often remarked upon, and, once observed, its peculiarity dismissed as if the combination of two such contradictory sources of criticism was so weird as to disqualify its objections. Subsequently, after 9/11, it has become noted that one of the elements of Western postmodern culture that particularly disgusts Islamic enemies is the inescapability of pornographic imagery seen even in advertisements for automobile tires and soft drinks. It is difficult to make the case against Islamic subjugation of women by defenders of Western ideals when the exploitation of women's bodies, and of sexuality in general, is so apparent in Western culture. The ideal of women as equals to men must first somehow be disentangled from the constancy of blatant pornography.[11] Despite the force of the earlier objections, the variety of sources of the criticism, and the ubiquity of the sexual imagery and innuendo itself the early reactions against pornography did not garner the attention they warranted. However, now that criticism of pornography and cultural sexualization has obtained the imprimatur of the mainstream media and become the object of scientific study, the effects of "pornification" can no longer be ignored.

Reconnecting Science and Religion in Postmodern Culture

If the cathedrals of medieval thought, wrought with an architectural sensitivity that combined all aspects of human experience and divinely revealed doctrine, are no longer available to us, what recourse does the postmodern experience offer by which to reconcile religion and science? The experience, rather than the thought, is paramount here because of the lack of an overarching philosophy. But if there is a paucity of intellectual structures, there are examples of how religion and science work in tandem in the postmodern social environment; the prior section used pornography as an example to illustrate this point.

The example of pornography makes the initial religious response resemble the canary in the mine—the bird that was brought into the dungeons of coal seams to detect the presence of poisonous gasses. The bird was expendable so that, once warned, the human beings could do the serious work of escape. To complete the analogy, the men in the mine can correspond to science, which does the real work of humanity as it solves its problems; the poisonous gasses to the presence of moral decay in the culture; and religion to a kind of early warning system which detects in advance social moral decay. However, this image denigrates religion in an unacceptable manner, for the detection of moral decay is not a sign of lack of intellectual rigor but a sign of ethical awareness of the kind that is systematically avoided in science. This includes social science even though it has the behavior of human beings as its principle object of research.

In the example of pornography, the religious objection is not, as the stereotype says, based on prudery and shameful guilt, but on the clear statements of doctrine that prescribe sexual behavior from a divine point of view. If wanton or promiscuous sexual activity is inherently wrong, so must those things be morally wrong which provoke or excite it, such as pornography. Furthermore, religious rules of social order often have the good of human beings as their direct purpose; as Jesus said, the law made for man, not man for the law. In the case, for example, of the commandment to "keep holy the sabbath" (Exodus, Chapter 31), while the putative intent is to worship the Lord in a proper manner by setting aside a day for him, the direct effect is to impose a day of rest after six days of labor. The elimination of the so-called "blue laws" in Massachusetts, those remnants of actual Puritan culture, was promoted as a release from religious strictures that have no place in the contemporary world. However, the workweek in Massachusetts is now seven days long, the seventh day is not a day of rest, but a day reserved for compulsive shopping. Stores and malls are serviced by a small army of retail workers, and so there is no break in the workweek and no rest from continual labor and frenetic activity. Applied to pornography, religious objections at first were characterized mainly as a threat to the First Amendment right of free expression, now appear to have been a prescient warning that the sexualization of our culture was taking place and that the results would not be good for our children or ourselves. By contrast, the scientific study of pornography is taking place only after the fact, in a manner approved by scientific protocols, but only after the damage has been done. Pornography serves as an example of the differences between how religion and science approach moral issues in the postmodern world but also as importantly provides a practical example of religion and science working together.

The relation of science to religion in this dynamic is not one of later strength replacing early weakness, or of positively defined knowledge replacing vague but compelling emotion. Science in this dynamic has its own weakness as well its peculiar strength of epistemological positivity. Its weakness is that scientific method deliberately prescinds from making value judgments, a strategy which enables scientific study to remain objective without regard to whatever philosophical or religious consequences its theories and reports of observed fact may have. Thus, teleology is withdrawn from biological explanation by evolution and the homeliness of the earth centered planetary system removed by the Copernican theory. The Copernican shift, it can be plausibly stated, merely offended Aristotelian professors of philosophy and astrological practitioners; it is more difficult to claim that the Darwinian shift was of equally little consequence for mechanical explanations of the ultimate origins of the human species does have an effect on human self-regard and the culture's understanding of the meaning of life. However, the situation is worse when the social sciences approach their phenomena for the effects are direct.

By prescinding from value judgments, social science can seem to imply that the human essence is no more than what empirical methods can study. The result can be seen in the succession of social scientists, including Hobbes, Comte, Marx, Feud, Skinner, and others, who provided a mechanical portrayal of human nature which categorically deprived humanity of its sense of being special. It is as if belief in the uniqueness of man is merely an irrational blockage to the fully scientific understanding of human nature—as if it were intellectually scandalous to anthropomorphize human nature. Does the researcher of the phenomena of crime not care about the effects of criminal activity on its victims? Is the researcher of hyperinflation unaware of the social and political destructiveness of runaway inflation? Now the concern of an individual social scientist may translate into a motivation for doing research into a particular kind of phenomena, but the value judgments about the effects of crime and of the persons who commit crimes is supposed to be independent of the methods and results of the study itself.

In reality, of course, what social scientists think about the phenomena they study and their judgments about them are frequently apparent. Their valuations sometimes are so powerful that they distort their purportedly objective findings. But the point is that religious doctrine supports the basis for value judgments about personal human behavior and social trends that can be supplied by no other elements in society or within human experience. With its lists of commandments that condemn certain behaviors as immoral, and its encouragement of actions that are defined as "good" religion gives a bright-line standard by which to judge human behavior and social trends and, thereby, provides a benefit to the culture at large. It may well be that religious standards are arguable or need to be refined according to the exact situation, but they are an essential part of the process of evaluation and legitimate forms of moral judgments. Indeed, without religious judgments it is hard to see how society can operate and what other social entity can or will provide them. As it happens, postmodern science hints at the possibility of a restoration of a teleological view of nature, but it is not so far developed or accepted as to provide a basis for cultural debate about the morality of new technological applications.[12] We are in a chaotic cultural situation in which philosophical systems of morality and reliance on calm discourse do not provide a basis for moral judgments. However, the more forceful remnant elements of postmodern culture do provide such a basis, namely, revealed religion and modern science. By attracting the support of large numbers of people, diverse elements of the population (including both religious believers and rationalists), and by providing a rational basis for the discernment of moral values and the means of their practical applications, religion and science can and do work together.

Notes

1.   Wolfhart Pannenberg, *An Introduction to Systematic Theology* (Grand Rapids, MI: Eerdmans, 1991), 44–49. Teilhard, *Phenomenon*, 291–294.
2.   John Caiazza, "The Arrival of Techno-secularism," *Modern Age* 44, 3 (Summer 2002), 208–16.
3.   *Boston Globe*, (March 8, 2005), 1.
4.   Nathan Glazer, "On Subway Graffiti in New York City," *Public Interest*, (Winter, 1979). This article is the seminal piece that has provoked much discussion and controversy.
5.   William Saletan, *Bearing Right: How Conservatives Won the Abortion War* (Berkeley: University of California Press, 2004).
6.   Steve Allen, *Vulgarians at the Gate* (Amherst, NY: Prometheus, 2001).
7.   Don Aucoin, "The Pornofication of America," *Boston Globe* (January 24, 2006), C1, C6.
8.   *New York Times*, "Children, Media and Sex: A Big Book of Blank Pages" (Jan. 31, 2006), in "Personal Health" section (on-line at www.nytimes.com).
9.   Nancy Kercheval (Bloomberg) "Research Links Media to Adolescent Sex," *Boston Globe* (April 4, 2006).
10.   American Psychological Association, 2007, "Report of the Task Force on the Sexualization of Girls Executive Summary." http://www.apa.org/pi/wpo/sexualizationsum.html.
11.   Appiah, *Cosmopolitanism*, 82, 83.
12.   Caiazza, "Natural Right and the Re-Discovery of Design," 273–309.

15

Conclusion: A Vision of Tolerable Order

The argument of this book has been somewhat complex. It has argued that the religion/science conflict is a real one, not a relic of the nineteenth century, and is taking place in contemporary culture, which has been termed "postmodern." However, no attempt has been made to resolve the conflict on a grand intellectual scale but rather to argue two major points. The first point (as argued in chapters three through ten), is that the conflict, while real and while seeming to favor science over religion, has lately turned to where the points made on both sides result in a wash—in equality in points of evidence and force of arguments—because science has entered a postmodern phase. As physicist-theologian John Polkinghorne put it, "we have all left the realm of knockdown argument behind."[1] These middle chapters present recent episodes or discoveries of contemporary science that bear on the relation of science to religion and attempt to deal with the conflict on the ground, that is, on the level of actual explanation and not an abstract philosophical level. Second, the argument is made that the conflict between religion and science is best dealt with in the context of their respective places in postmodern culture, since postmodern culture has a chaotic and unhealthy aspect and that both science and religion are required to make necessary repairs. The postmodern world may well reply that it does not need the repairs offered by religion and science, that the imposition of master narratives upon society has had destructive effects, are false, and exist merely as a pretext for maintaining power by elite elements. It is this last general point that needs to be answered to complete the full argument of the book.

The first point to be made in response is that the embrace of chaos and the antinomianism of postmodern culture expressed in sexual latitudinarianism, libertarian politics, attitudinal anti-authoritarianism, and a kind of menacing sloppiness in dress and manners is a reaction to those destructive features which characterized the twentieth century. The twentieth century realized two aspects of social conformity at their worst: whole nations organized with great efficiency for warfare with all of the industries directed at the manufacture of weapons and the men subject to universal conscription and the imposition on some nations of extremely cruel totalitarian regimes including Fascism, Communism, and Nazism. The effects of this intense degree of social conformity, which was made possible by the application of new technologies were a series of world wars among nations, many revolutions within nations, and untold suffering and death for tens of millions of individuals. The rejection of the possibility of a set of publicly acknowledged rules or ethical constraints in the mores of the culture or the law of nations which is characteristic of postmodern culture is based on fear and loathing of the conformity enforced by governments in the twentieth century. The calamities provoked by nations under the most intense social pressure is a powerful memory and one which will affect postmodern culture for a very long time to come; it is a memory which will, unfortunately, skew the understanding of the role of publicly maintained behavioral rules.

The result of this memory is the rejection of the sense of moral order, of the very idea that, in some fashion or other, the universe makes sense. Confidence that the universe makes ultimate sense seems almost to be an insult to the victims of twentieth-century warfare and the Holocaust such that belief in an objective order of reality becomes a sign of triviality—that the believer in the possibility of moral or physical order just "doesn't get it." Thus, the postmodern sensibility literally sees no difference between the Ten Commandments and the orders of a Nazi camp officer; they are both unjustified restrictions of personal freedom. In the light of this degree of rejection, any suggestion that a sense of order is required for the "repair" of postmodern culture will have to be made with a great deal of sensitivity.

The virtue of suggesting that *both* religion and science should be more forcefully recognized within postmodern culture and made essential elements of the postmodern worldview is that while they are "master narratives," they are separate and in some degree of opposition. There is not a uniformity of principle between them except that they both attempt to make sense of the universe. By accepting religion and science and reinstituting them in the culture, then, the postmodern world would not subject itself to a new kind of totalitarian control but to the possibility of order and structure in our understanding of an otherwise chaotic universe. As for the widely accepted notion taken from French postmodernists that master narratives and logic itself are attempts to control society on behalf of elites, it is perhaps enough to point out that such an assertion is self-contradictory. Such analysis itself depends upon the use of logic and by accepting it, the hearer of such an analysis accepts also the imposition of a new elite. This is not the white-shoe rich or the old boys' network, but a new elite of social critics and college professors who, notoriously, cannot be voted out of office or fired for incompetence. In any case, one of the actual elements of the postmodern world is that while there are elites (social philosophers including Pareto and Burke have pointed out elites are inevitable in society), there are many elites and not just a single party or class that controls industry, finance, opinion, research, education, entertainment, and, yes, science and religion. The guarantee of personal freedom so highly regarded in postmodern culture is best found not in a frenzied search and destroy mission against elites, traditions, or narratives, but in the fact that there are many elites and we are free to adjust ourselves, our needs, desires, goals, and ideas to fit them as we, as individuals, see fit.

As experienced, the primary aspect of living in the postmodern world is dissociation—the lack of connection of its component elements–which is to say that it lacks an inner sense of order and thus feels to be essentially incoherent. That, it may be said, is the price you pay for living in a truly world culture—i.e., one in which all of the cultures and all of the classes of people of the world are recognized on their own terms. The problem remains, however, that one cannot live in confusion forever, particularly when action and the daily process of living require decision and evaluation; thus we cannot escape the need to make moral decisions. For without commonality, one is left with the assertion of one's own ego and the demand upon society that it fulfills one's personal desires; the means of mediating the interpersonal and social conflicts that result is presumed to lie in technology and tolerance. However, the technology rests in large part on cheap energy and the oil is running out—and tolerance is now stretched beyond the breaking point of accepting deadly attacks upon one's own people. A refreshing sense of the possibility of reality, of knowledge of an ethical order, and physical structures beyond one's own ability to change things is required. Interconnectivity and association cannot be imposed by social authority; they must come from within individuals and smaller communities, including families. The fact that we are in the midst of creating a world culture makes the resolution of the inner incoherence of the postmodern worldview all the more important.

Here religion and science provide the means of resolution. Religion in the postmodern version which excludes none and prefers none seems a preferable ideal, but it is a concept of religion without definition and ultimately without moral force. Reliance on the religious precepts of Christianity primarily, but also Judaism and Islam, and inclusion of them within public discourse in the West will not put off other cultures but act rather to reassure them that in postmodern culture religion has a revered place. The elimination of the Abrahamic religions from our public square will not convince religious believers in other cultures that we will respect theirs. Religion is essential as the guarantor of ethical normality and self-discipline of the civil order and perhaps ultimately as the guarantor of the rationality of the universe.

Science and scientific method are resented in the postmodern world in the same manner as parental orders are resented by young children, as if they were the arbitrary imposition of someone else's will. Yet reality makes its demands nonetheless, so much so that Freud made it a principle of psychology. The reality principle harshly reduces the search for utopia from an achievable political agenda to a feckless dream by imposing knowledge of the permanent limits of human nature described by the physical, biological, and social sciences. Science as the guarantor of a persistent reality is a means to psychological health for individuals and postmodern culture at large. Science also provides knowledge of the physical laws that are the basis of the technological means of dealing with reality in a prudential manner. Because they are unavailable from any other cultural source, without the repairs offered by religion and science it is possible to predict that postmodern culture will continue its descent into intellectual confusion and moral chaos. As it becomes the universal culture of the world, mankind will be in a position of hastening to live as quickly and pleasurably as possible before cultural decomposition overcomes the ability of technology to deal with the consequences.

The problem with which this book began, the conflict between religion and science as expressed by the Jesus and Darwin fishes, has not been overcome by means of a master narrative corresponding to a medieval cathedral that combines various conflicting elements in precarious but permanent stone-secure balance. Rather, the conflict is understood to persist, but the continued presence of both sides of the conflict is required because of the view of comprehensive reality that each provides— albeit in very different ways. However, the interaction of religion and science is now being accomplished on the practical level, which gives us confidence that resolution is eventually possible on the intellectual and cultural levels as well. In the meantime, the vision of comprehensible reality, interconnection, and coherence in physical, moral, and divine dimensions is required in the creation of the new, postmodern world—a vision of tolerable order. Science demonstrates to us that truth exists, and religion reveals that the truth will make us free.

Notes

1.   John Polkinghorne, *The Way the World Is* (Grand Rapids, MI: Eerdmans, 1983), 6.

Bibliography

Allen, Steve. *Vulgarians at the Gate*. Amherst, NY: Prometheus, 2001.

American Psychological Association, 2007, "Report of the Task Force on the Sexualization of Girls Executive Summary," http://www.apa.org/pi/wpo/sexualizationsum.html.

Appiah, Kwame Anthony. *Cosmopolitanism*. New York: W. W. Norton and Company, 2006.

Appleyard, Bryan. *Understanding the Present: Science and the Soul of Man*. New York: Doubleday, 1992.

Asimov, Isaac. *It's been A Good Life,* edited by Janet Jeppson Asimov. Amherst, NY: Prometheus, 2002.

Aucoin, Don. "The Pornofication of America," *Boston Globe* (Jan. 24, 2006): C1, C6.

Saint Augustine. *Confessions*, translated by F.J. Sheed. New York: Sheed and Ward, 1943.

———. *On Christian Teaching,* translated by R. P. H. Green. New York: Oxford University Press, 1999.

Barbour, Ian G. *Religion and Science: Historical and Contemporary Issues*. San Francisco: HarperCollins, 1997.

Barrow, John and Tipler, Frank. *The Cosmic Anthropic Principle*. New York: Oxford University Press, 1998.

Behe, Michael. *Darwin's Black Box*. New York: Free Press, 1996.

Bird, Kay and Sherwin, Martin J. *American Prometheus*. New York: Alfred A. Knopf, 2005.

Boyer, Pascal. *Religion Explained*. New York: Basic Books, 2001.

Brody, Jane E "Children, Media and Sex: A Big Book of Blank Pages," *New York Times* (January 31, 2006): "Personal Health" section.

Bronowski, Jacob. *The Ascent of Man*. Boston: Little Brown, 1973.

Bunch, Bryan. *Handbook of Current Science and Technology*. Detroit: Gale, 1996.

Burtt, Edwin A. *The Metaphysical Foundations of Modern Science*. Garden City. NY: Doubleday, 1954.

Butterfield, Herbert. *The Origins of Modern Science 1300–1800*. New York: Free Press, 1957.

Caiazza, John. "Review of Appleyard." *Modern Age* 37, 3 (Spring, 1995): 257–260.

———. "Natural Right and the Re-Discovery of Design in Contemporary Cosmology," *The Political Science Reviewer* XXV (1996).

———. *Can Religious Believers Accept Evolution?* Huntington, NY: Troitsa Books, 2000.

———. "The Arrival of Techno-secularism," *Modern Age* 44, 3 (Summer, 2002): 208–216.

——. "Athens, Jerusalem and the Arrival of Techno-secularism," *Zygon* 40, 1 (March, 2005).

——. "The Athens/Jerusalem Template and the Techno-secularism Thesis," *Zygon* 51, 2 (June, 2006): 235–248.

Capra, Fritjof. *The Tao of Physics* (third edition). Boston: Shambhala, 1991.

Collingwood, R.G. *The Idea of Nature.* New York: Oxford University Press, 1945.

Cosmides, Leda and Toohy, John. "Evolutionary Psychology: A Primer," Center for Evolutionary Psychology online, www.psych.ucsb.edu/research/cep/primer.html.

Crick, Francis. *Life Itself.* New York: Simon and Schuster, 1981.

Darwin, Charles. *The Origin of Species.* New York: Modern Library, (no date).

——. *The Descent of Man.* New York: Modern Library, (no date).

Davies, Paul. *God and the New Physics.* New York: Simon and Schuster, 1984.

Dawkins, Richard. *The Selfish Gene.* Oxford: Oxford University Press, 1989.

——. *River Out of Eden.* New York: Basic Books, 1995.

——. *The God Delusion.* Boston: Houghton Mifflin, 2006.

Dennett, Daniel. *Darwin's Dangerous Idea.* New York: Simon and Schuster, 1995.

——. *Breaking the Spell.* New York: Penguin, 2006.

Derrida, Jacques. *Rogues: Two Essays on Reason.* Stanford, CA: Stanford University Press, 2005.

DeSantillana, Georgio. *The Crime of Galileo.* Chicago: University of Chicago Press, 1959.

Duhem, Pierre. *The Aim and Structure of Physical Theory*, translated by P.P. Wiener. Princeton, NJ: Princeton University Press, 1954; reprint New York: Atheneum, 1977.

Einstein, Albert and Infeld, Leopold. *The Evolution of Physics.* New York: Simon and Schuster, 1961.

Falk, Dan. *Universe on a T-shirt.* New York: Arcade Publishing, 2005.

Ferguson, Kitty. *The Fire in the Equations.* Grand Rapids, MI: Eerdmans Publishing Company, 1995.

Feyerabend, Paul. *Against Method.* New York: Verso Books, 1988.

Feynman, Richard. *The Character of Physical Law.* New York: Modern Library, 1994.

——. *Surely You're Joking, Mr. Feynman.* New York: W. W. Norton and Company, 1997.

Fisher, R. A. "Has Mendel's Work Been Rediscovered?" *Annals of Science* 1 (1936).

Foucault, Michel. *The Birth of the Clinic: An Archeology of Medical Perception.* New York: Vintage Books, 1975.

Fuller, Steven. *Thomas Kuhn: A Philosophical History For Our Times.* Chicago: University of Chicago Press, 2000.

Galileo, Galilei. *Dialog Concerning the Two Chief World Systems,* translated by S. Drake. Berkeley: University of California Press, 1967.

———. "Letter to the Princess Christina." In *The Galileo Affair: A Documentary History,* edited and translated by M. A. Finocchiaro. Berkeley: University of California Press, 1989.

Galison, Peter. *Einstein's Clocks, Poincare's Maps.* New York: W. W. Norton and Company, 2003.

Gillispie, Charles. *The Edge of Objectivity.* Princeton, NJ: Princeton University Press, 1960.

Glazer, Nathan. "On Subway Graffiti in New York City," *Public Interest* (winter, 1979).

Goldberg, Carey. "Crossing Flaming Swords over God and Physics," *New York Times* (April 20, 1999): D-5.

Gould, Stephen J. *The Panda's Thumb.* New York: W. W. Norton and Company, 1980.

———. "Sex, Drugs, Disasters and the Extinction of Dinosaurs." In *The Flamingo's Smile.* New York: W. W. Norton and Company, 1985.

———. *Wonderful Life.* New York: W. W. Norton and Company, 1989.

———. *Rocks of Ages: Science and Religion in the Fullness of Life.* New York: Ballantine, 1999.

Gross, Paul R. and Levitt, Norman. *Higher Superstition: The Academic Left and its Quarrels with Science.* Baltimore, MA: Johns Hopkins University Press, 1998.

Hamer, Dean. *The God Gene: How Faith Is Hardwired Into Our Genes.* Garden City: Doubleday, 2004.

Harkin, Gregg and Kevles, Daniel J. "The Oppenheimer Case: An Exchange," *New York Review of Books* 51, 5 (March 24, 2004).

Hart, Jeffrey. "What is the 'West'?" *Modern Age* 47, 4 (Fall, 2004): 362–366.

Haught, John F. "Science and Scientism: The Importance of a Distinction," *Zygon* 40,2 (June, 2005), 363–368.

Hawking, Steven. *A Brief History of Time.* New York: Bantam Books, 1988.

Hertz, Heinrich. *The Principles of Mechanics Presented in a New Form.* New York: Dover, 1956.

Himmelfarb, Gertrude. *Darwin and the Darwinian Revolution.* New York: Anchor Books, 1962.

Horgan, John. *The End of Science.* Reading, MA: Addison-Wesley, 1996.

Horwich, Paul, ed. *World Changes.* Cambridge, MA: MIT Press, 1993.

Hoyle, Fred. *Nicolaus Copernicus.* New York: Harper and Row, 1973.

Jastrow, Robert. *God and the Astronomers.* New York: Warner, 1978.

Jones, W. T. *A History of Western Philosophy, Vol. III Hobbes to Hume.* New York: Harcourt World, 1969.

Kant, Immanuel. "The Metaphysical Foundation of Morals." In *The Philosophy of Kant,* translated by C. J. Friedrich. New York: Modern Library, 1949.

Keller, Evelyn Fox. *A Feeling for the Organism.* New York: W. H. Freemen, 1993.

Kercheval, Nancy. "Research Links Media to Adolescent Sex," *Boston Globe* (April 4, 2006).

Koyre, Alexendre. *From the Closed World to the Infinite Universe.* Baltimore, MA: Johns Hopkins University Press, 1957.

Kuhn, Thomas. *The Copernican Revolution.* Cambridge, MA: Harvard University Press, 1957.

———. *The Structure of Scientific Revolutions* (second edition). Chicago: University of Chicago Press, 1970.

Lakatos, Imre and Musgrave, Alan, eds. *Criticism and the Growth of Knowledge.* Cambridge: Cambridge University Press, 1970.

Larsen, Rebecca. *Oppenheimer and the Atomic Bomb.* New York: Franklin Watts, 1988.

Lewis, C.S. *Out of the Silent Planet.* New York: Scribner, 1938.

Lourie, Richard. *Sakharov: A Biography.* Hanover, NH: Brandeis University Press, 2002.

Lyotard, Jean-Francois. *The Postmodern Condition.* Manchester: Manchester University Press, 1984.

Manuel, Frank E. *A Portrait of Isaac Newton.* Washington, DC: New Republic, 1968.

McMillan, Pricilla J. *The Ruin of J. Robert Oppenheimer and the Origin of the Modern Arms Race.* New York: Viking, 2005.

Milton, John. *Paradise Lost.* New York: Penguin Classics, 2003.

Pannenberg, Wolfhart. *An Introduction to Systematic Theology.* Grand Rapids, MI: Eerdmans Publishing Company, 1991.

Pascal, Etienne. *The Pensees* (trans. J.M. Cohen). Baltimore: Penguin Books, 1961.

Polkinghorne, John. *The Way the World Is.* Grand Rapids: Eerdmans Publishing Company, 1983.

———. *Belief in God in an Age of Science.* New Haven, CT: Yale University Press, 1998.

Popper, Karl. *Conjectures and Refutations.* New York: Harper and Row, 1965.

Radford, Tim. "Science cannot provide all the answers," *The Guardian* (Sept. 4, 2004).

Rhodes, Richard. *The Making of the Atomic Bomb.* New York: Simon and Schuster, 1995.

Roland, Wade. *Galileo's Mistake.* New York: Arcade Publishing, 2003.

Rorty, Richard. *Philosophy and the Mirror of Nature.* Princeton, NJ: Princeton University Press, 1979.

———. *Contingency, Irony and Solidarity.* New York: Cambridge University Press, 1989.

Rosenberg, Alexander. *Instrumental Biology or the Disunity of Science.* Chicago: University of Chicago Press, 1994.

Sagerstrale, Ullica. *Defenders of the Truth.* New York: Oxford University Press, 2000.

Saletan, William. *Bearing Right: How Conservatives Won the Abortion War.* Berkeley: University of California Press, 2004.

Sapp, Jan. "The Nine Lives of Gregor Mendel." In *Experimental Inquiries,* edited by H. E. LeGrand. Kluwer Norwell, MA: Academic Publishers, 1991.

Schilpp, Paul A., ed. *Albert Einstein: Philosopher-Scientist; Library of Living Philosophers Vol. VII.* Chicago: Open Court, 1949.

Schroeder, Gerald L. *The Science of God.* New York: Broadway Books, 1997.

Snow, C. P. *The Two Cultures.* Cambridge: Cambridge University Press, 1993.

Sobel, Dava. *Galileo's Daughter.* New York: Walker, 1999.

Strauss, Leo. *Natural Right and History.* Chicago: University of Chicago Press, 1953.

Teilhard de Chardin, Pierre. *The Phenomenon of Man,* translated by B. Wall. New York: Harper and Row, 1959.

Thorndyke, Lynn. *History of Magic and Experimental Science.* 1923.

Wartofsky, Marx. *Conceptual Foundations of Scientific Thought.* New York: Macmillan, 1968.

Weinberg, Steven. *Dreams of a Final Theory.* New York: Pantheon, 1992.

———. "The Revolution That Didn't happen," *New York Review of Books* XLV, 15 (1998).

Wheelright, Philip. *The Presocratics.* Indianapolis: Bobbs-Merrill, 1982.

Wilson, Edward O. *Sociobiology: The Abridged Edition.* Cambridge, MA: Harvard University Press, 1980.

———. *On Human Nature.* Cambridge, MA: Harvard University Press, 1978.

———. *Consilience.* New York: Vintage, 1998.

Index